LOCOMOTION PAPERS	LP244

THE
WHITBY – LOFTUS LINE

by
Michael A. Williams

With not a passenger in sight, 'L1' class No. 67754 waits at Whitby (West Cliff) station with a Scarborough-bound train at the end of March 1958, five weeks before the end. *I.S. Carr*

THE OAKWOOD PRESS

First published 2012 as the Whitby-Loftus Line , a history
by the Jet Coast Development Trust.

This revised and expanded edition © Michael A. Williams, 2019.

ISBN 978-0-85361-542-2

Printed by
Blissetts, Shield Drive, West Cross Industrial Park, Brentford, TW8 9EX

Crossing Sandsend Beck above the roofs of the hotels and houses.
J.W. Armstrong/Armstrong Photographic trust

Published by
The Oakwood Press, 54-58 Mill Square, Catrine, KA5 6RD
01290 551122   www.stenlake.co.uk

Contents

Introduction		5
One	A Spectacular Failure	7
Two	Early Maps and Plans	33
Three	A Difficult Year in the History of the Whitby, Redcar and Middlesbrough Union Railway	39
Four	The Viaducts and Tunnels of the Whitby – Loftus Line	53
Five	A Brief Financial History of the Line from 1897 to 1940	93
Six	The Importance of Fieldwork in Researching the History of the Whitby – Loftus Line	115
Seven	The Suez Specials	133
Eight	Closing a Line Before Beeching: the End of the Whitby – Loftus Line	139
Nine	The Railway in the Imagination	155
Ten	Whitby's Third Station	157
Eleven	The Station that Never Was	161

Appendices

One	Transcript of the T. E. Harrison memorandum of 14th November, 1883	165
Two	Total receipts for all stations for 1897-1907	167
Three	Deepgrove tunnel derailment. LNER (NE area). Report of accident	167
Four	Report of Deepgrove tunnel derailment accident.	168
Five	*Whitby Gazette* article of 5th August, 1927: 'Excursion train off the line'	169
Six	Occupations of Whitby, Redcar & Middlesbrough Union Railway Debenture holders	170
Seven	The Railway Inspectorate: Inspectors' Reports (1883)	171

Bibliography	175
References	179
Index	187

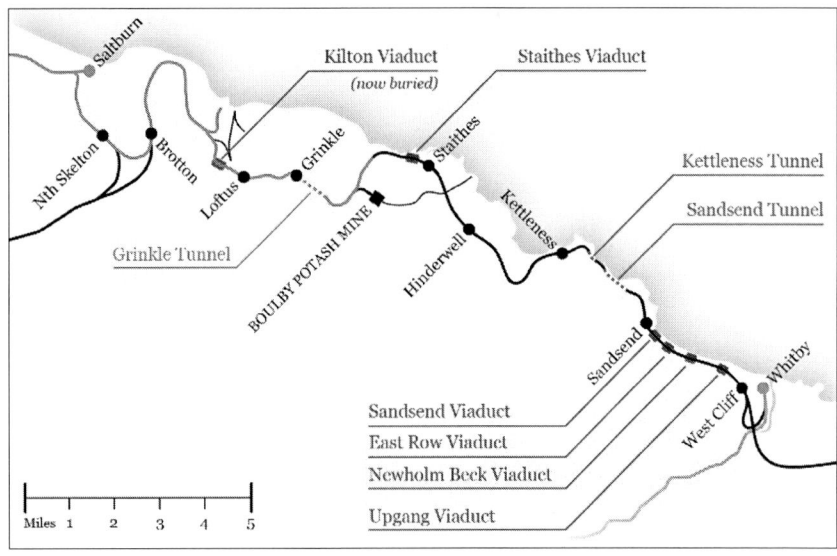

The Whitby – Loftus line. *Forgotten Relics of an Enterprising Age*

Railways in the North Yorkshire area (the line seen in a wider perspective).

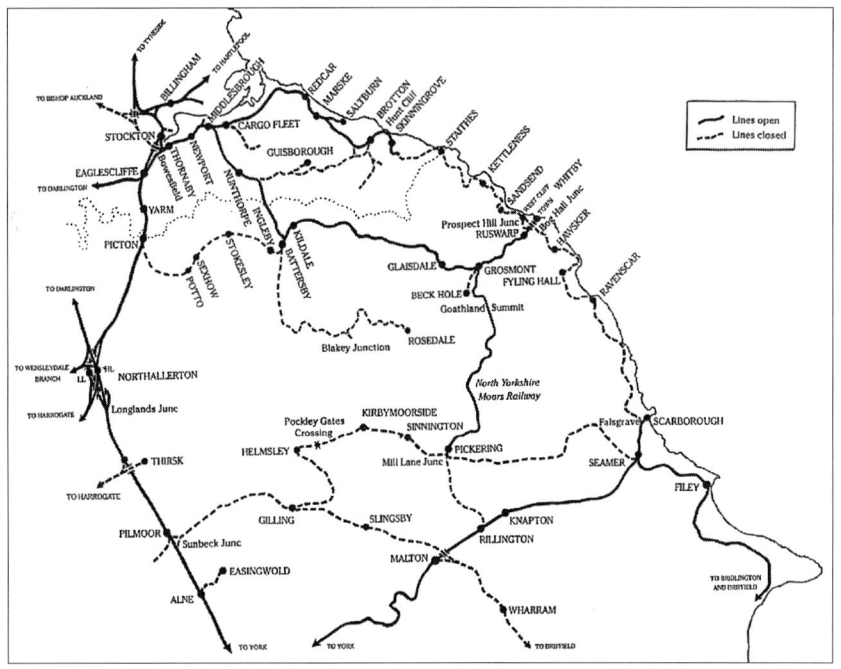

Introduction

Saturday 6th March, 1965 was the darkest day in Whitby's railway history with the closing of two lines simultaneously (Whitby - Scarborough and Whitby - Malton). So important were these closures to Whitby that it has meant that the earlier closing of another Whitby line, to Loftus, is now almost forgotten. May 2018 marked the 60th anniversary of the line's closure. The last trains to Whitby along this line ran on Saturday, 3rd May. The line ran from Whitby Town station, via West Cliff station, Sandsend, Kettleness, Hinderwell, Staithes, and Loftus. It offered the most spectacular and dramatic ride in the north of England travelling, as it did, over viaducts, through tunnels, along the edge of precipitous cliffs, giving panoramic views of sea, sand, valley, bay, abbey, castle, wood and – occasionally – shipwreck. The line cost over £50 million to build (in today's money) and took 12 years. Before it opened the line was inspected by the Board of Trade and five times it was denied permission to open because of perceived 'danger to the public'. The Tay Bridge disaster of 1879 meant that all the five viaducts had to be strengthened and, in the case of Staithes, a unique wind gauge attached.

The line lasted less than 75 years: it never paid its way, the maintenance costs of the Deepgrove tunnel alone were immense, and as soon as it became possible, British Railways hurriedly closed the line to avoid more costly expenditure. By the end, less than one person a day took the train at Sandsend. Nowadays the remains of the line can be followed for much of the 16 miles between Whitby and Loftus and, in fact, the last four miles were re-opened in 1974 to serve the Boulby potash mine.

Though little-used for most of the year, the line was busy during the six weeks or so of the summer season. The winter timetable in the last months of the line's existence offered two trains a day from Scarborough and one from Whitby (Town) to Loftus and Middlesbrough. There were just two trains returning to Whitby and Scarborough. It was known since November 1957 that the line would close in early May and, during the last weeks, the trains – such as they were – were packed, for the line was loved by locals and visitors alike. The last day was sunny, bright, the sea was calm, and there was hardly any wind. The *Whitby Gazette* reporter went to Sandsend station to witness the passing of the last trains and wrote one of the finest descriptions about the closure of a rural branch line (see *Whitby Gazette* of 9th May, 1958). Today the viaducts have gone, the rails have gone, the tunnels are in a state of incipient collapse and much of the trackbed has been swept away by road improvements, building and farming. For those who travelled on the line the memories remain, for those who did not the following episodes in the history of the line will, it is hoped, give an insight into a brief, but important era in Whitby's railway history.

A Whitby-bound train crosses Eastrow viaduct during the early spring of 1957. Note the proximity of the line to the beach. *N Cholmondeley collection*

CHAPTER ONE

A SPECTACULAR FAILURE

Recent research on the line begins in 1993 with R. J. Irving's article in the *Journal of Transport History* which, *inter alia*, described the Whitby – Loftus line as a 'spectacular failure.'[1] Irving's rather harsh analysis of the line was modified by the present author's article in the *Journal of the Railway and Canal Historical Society* which drew the conclusion that Irving, though correct in many respects, overlooked the short periods of successful financial operation of the line and that, while it was a failure, the description of 'spectacular' was too extreme.[2] This latter article briefly illustrated the main argument of the full-scale history of the line, published in 2012.[3] However, whatever the degree, it cannot be denied that the line's 75 year existence was one of financial difficulty. Indeed, the financial history of the line can be summed up in one graph which shows the rise and dramatic decline of passenger bookings on the line between 1885 and 1940.

However, the phrase 'spectacular failure' can be read differently; part of the argument of this book is that the line should not necessarily be seen solely in financial terms but as an artefact which offered to its

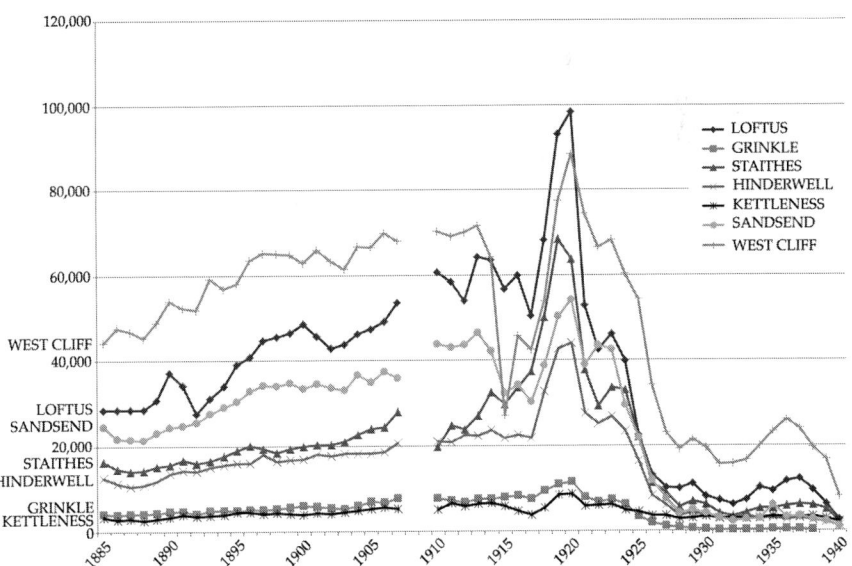

Passengers booked (all stations) 1885-1940. This graph summarizes perfectly the fortunes of the line during its 75 year lifetime.

travellers spectacular views of the North Yorkshire coast which no other form of transport has since equalled. Although occasionally touched upon in other railway histories, the concept of the line in the landscape as an added element of beauty is not one regularly considered. Many branch lines offered just that: man-made beauty which merged with the natural glory of the countryside through which it passed. The photographs of the line offer a confirmation of that argument. This idea is perhaps best exemplified when one considers the scene after a railway has been abandoned and destroyed.

**The route of the line
(with emphasis upon the railway in the landscape)**

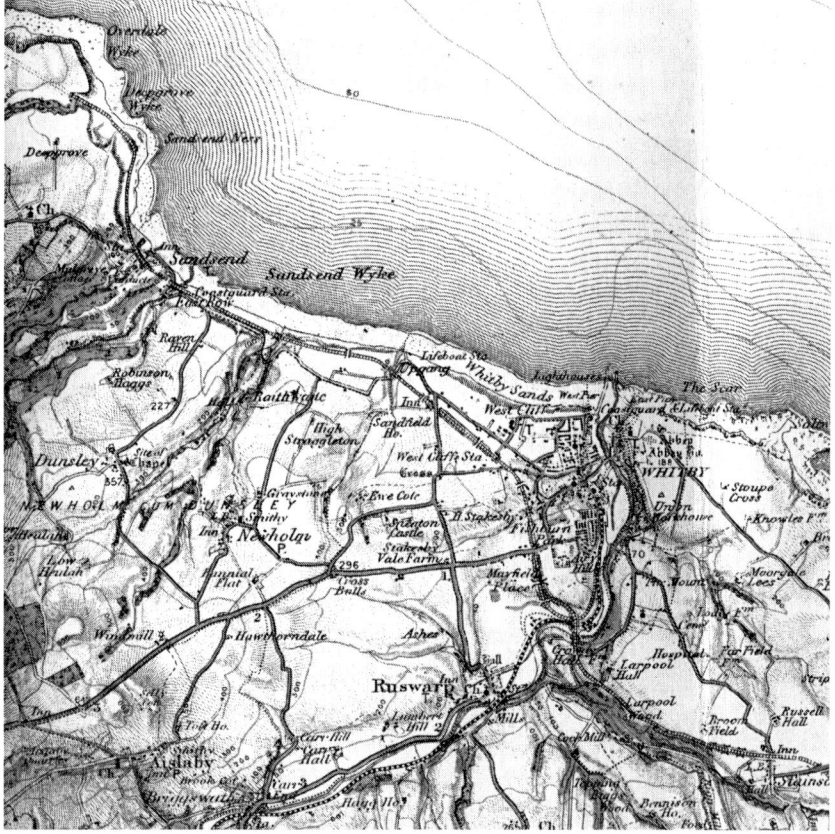

The line and its environs in 1904, from Whitby (Town) to Deepgrove tunnel. Ordnance Survey 1904 one inch to the mile coloured edition.

A SPECTACULAR FAILURE 9

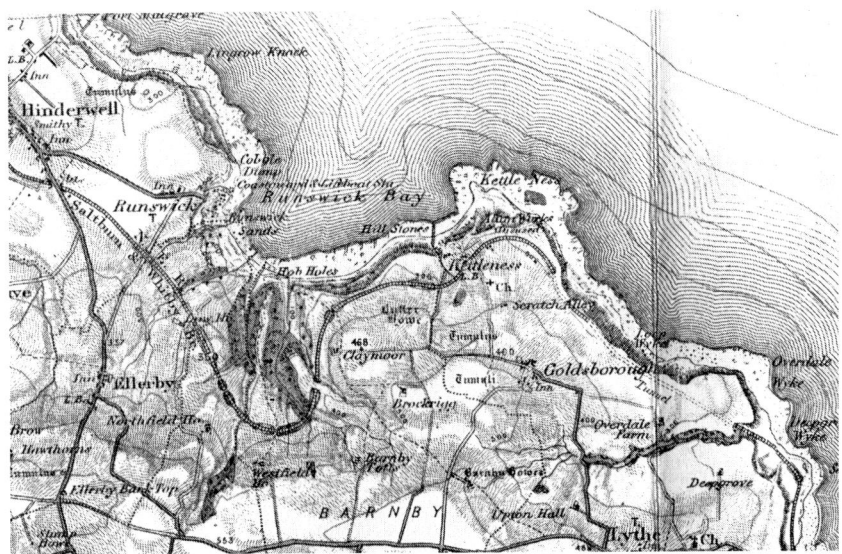

The line and its environs in 1904, from Deepgrove tunnel, to Hinderwell.

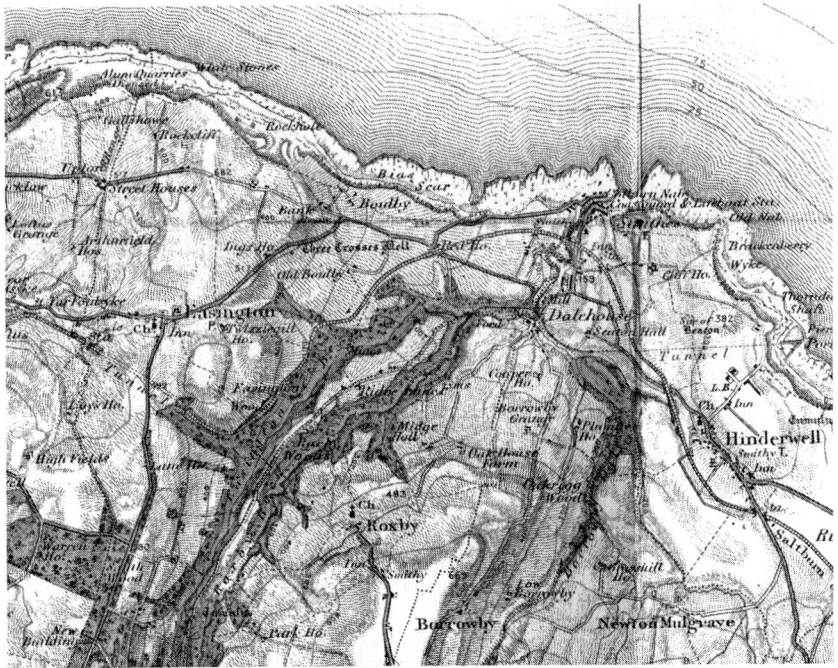

The line and its environs in 1904, from Hinderwell to Grinkle.

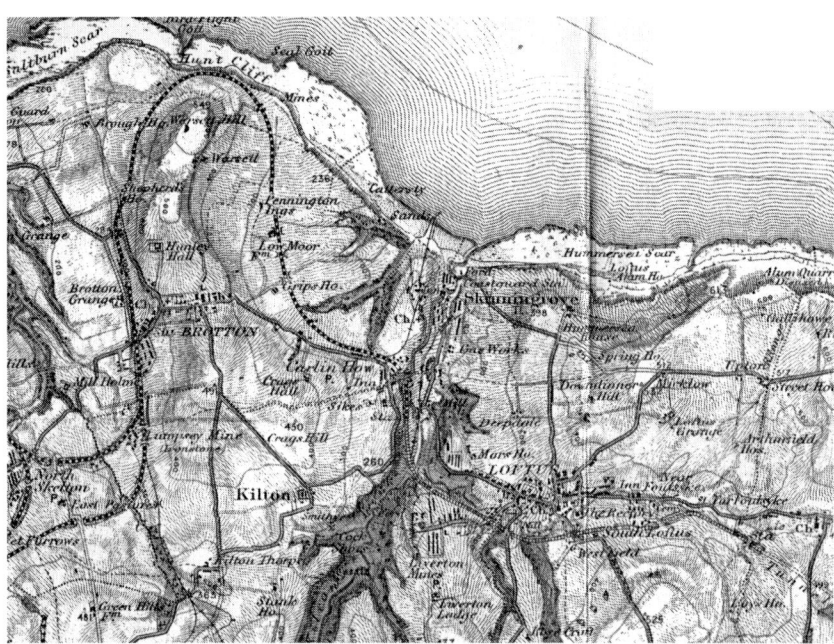

The line and its environs in 1904, from Grinkle to Loftus and on to North Skelton Junction.

The line in detail from Whitby (Town) to Whitby (West Cliff). O. S. map (7th ser., 1955) showing Whitby (Town) and Whitby (West Cliff) stations.

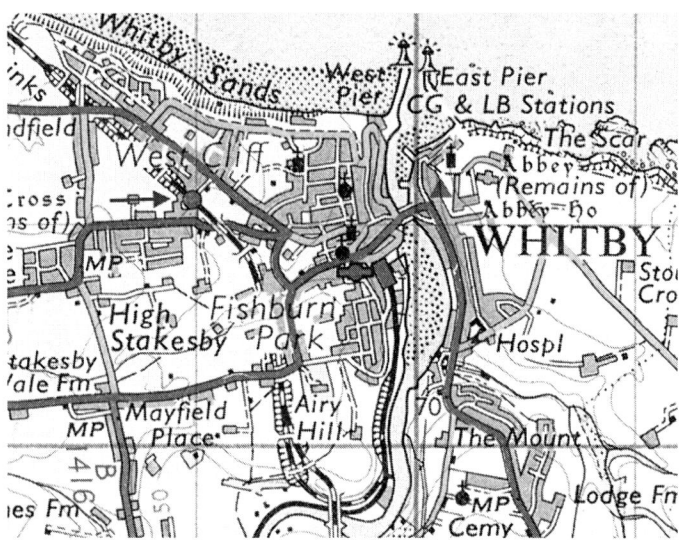

The line between Whitby (Town) and Whitby (West Cliff)

Whitby (Town) station interior in the early fifties. This is how the station would have looked for almost all the 75 years existence of the line. *British Rail*

The railway in the landscape. Railway infrastructure near Whitby (Town) (probably late twenties/early thirties). *Author's collection*

Bog Hall Junction; the beginning of the Whitby-Loftus line (*left*). *K. Hoole*

A light engine on the climb from Bog Hall Junction to Prospect Hill Junction seen from the top of Larpool viaduct. *M. Mensing*

A SPECTACULAR FAILURE 13

A Whitby (Town)-bound train coming down the bank from Prospect Hill Junction in the summer of 1957. *M. Mensing*

Prospect Hill Junction with Scarborough-bound train in August 1957. *A.M. Ross*

The line between Whitby (West Cliff) and Sandsend

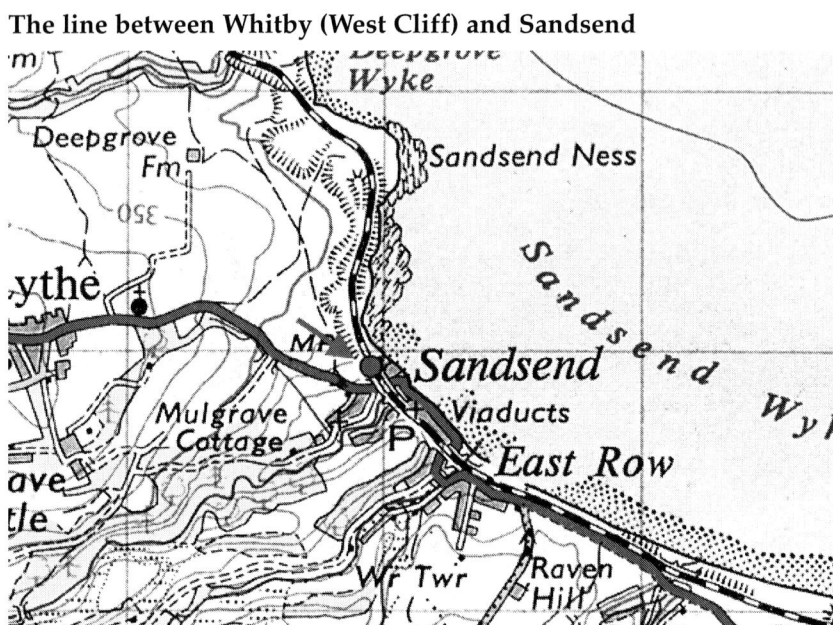

The line in detail from Whitby (West Cliff) [just off the map to the right] to Sandsend. O.S. map (7th ser., 1955).

Entering Whitby (West Cliff) station from the direction of Prospect Hill. *K. Field*

Whitby from the headland above Sandsend, George Weatherill [1810-90]. This seems to be the earliest representation of a train on the Whitby-Loftus line and was painted between 1884 (the first summer of the line) and (probably) 1889. This viewpoint (reached from a footpath from Lythe Bank) was also used by J. W. Armstrong over 70 years later.

Newholm viaduct between Whitby (West Cliff) and Sandsend.
J.W. Armstrong/Armstrong Photographic Trust

Ivatt class '4MT' No. 43074 crosses over Eastrow viaduct with a Whitby - Middlesbrough train, c. 1957. *J.W. Armstrong/Armstrong Photographic Trust*

The line at Eastrow (probably 1940s). *Author's collection*

A SPECTACULAR FAILURE 17

Summertime at Sandsend with a Middlesbrough-bound train crossing the viaduct.
J.W. Armstrong/Armstrong Photographic Trust

The railway in the landscape (Sandsend).
J.W. Armstrong/Armstrong Photographic Trust

The line between Sandsend and Deepgrove Tunnel

Looking south from the headland above Sandsend, a similar vantage point to George Weatherill's for his painting. *J.W. Armstrong/Armstrong Photographic Trust*

A wintry scene as a Middlesbrough-bound train is about to enter Deepgrove tunnel.
I.S. Carr

The line between Deepgrove Tunnel and Kettleness

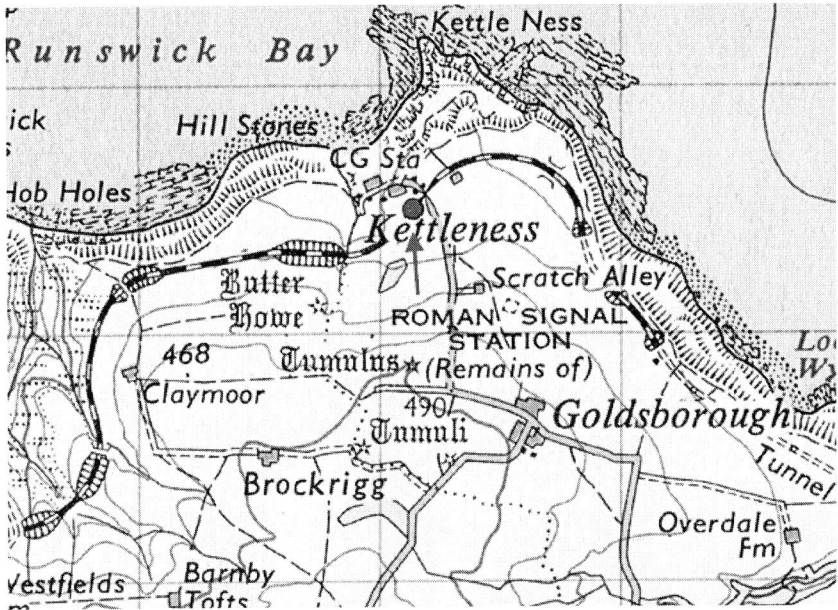

The line in detail from Deepgrove tunnel to Kettleness and onwards towards Hinderwell O. S. map (7th ser., 1955).

The railway in the dramatic landscape between Deepgrove and Kettleness tunnels.
J.W. Armstrong/Armstrong Photographic Trust

A Whitby-bound train about to enter Kettleness tunnel. *K. Hoole*

View of Kettleness station. *J.W. Armstrong/Armstrong Photographic Trust*

The line between Kettleness and Hinderwell

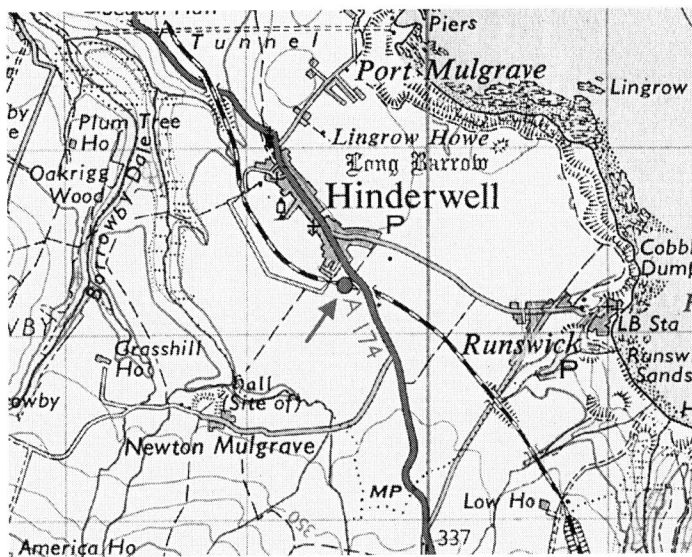

The line in detail from Kettleness [just off the map] to Runswick and Hinderwell; O. S. map (7th ser., 1955) showing Hinderwell station and the site of the proposed Runswick station.

Last day at Kettleness. *J.W. Armstrong/Armstrong Photographic Trust*

Kettleness station (*right*) in the very early years of the 20th century
Author's collection

Runswick Bay forms a dramatic backdrop for a Whitby-bound train passing the Kettleness distant signal. *C.C. Cobb*

A SPECTACULAR FAILURE

The last train north; looking towards Ellerby Lane from Hinderwell bridge.
A.K. Lamballe

The line from Hinderwell to Staithes

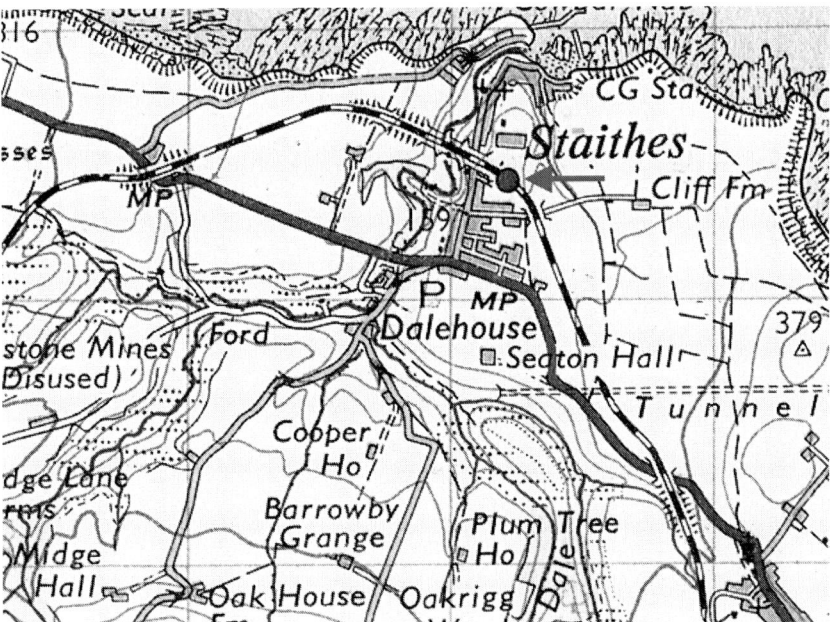

The line in detail from Hinderwell to Staithes; O. S. map (7th. ser., 1955) showing Staithes station.

Hinderwell station before the construction of the down platform.

Author's collection

Last train ever at Hinderwell (3rd May, 1958).

A.K. Lamballe

A Middlesbrough-bound train enters Staithes station.   *N. Cholmondeley collection*

A Whitby-bound train crosses Staithes viaduct.
J.W. Armstrong/Armstrong Photographic Trust

The line between Staithes and Grinkle

The last train on the line crosses the long-demolished bridge over the Easington-Staithes road. *A.K. Lamballe*

The line at Grinkle

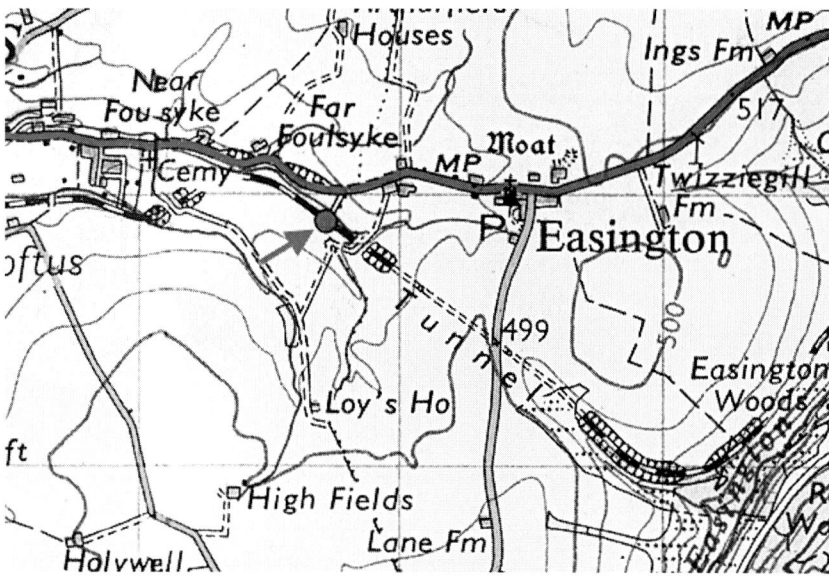

The line in detail at Grinkle (Easington); O. S. map (7th ser., 1955) showing Grinkle station.

A SPECTACULAR FAILURE

The remains of Grinkle station looking towards Loftus.
J.W. Armstrong/Armstrong Photographic Trust

A Middlesbrough-bound train leaves Grinkle (Easington) tunnel.
J.W. Armstrong/Armstrong Photographic Trust

The line at Loftus

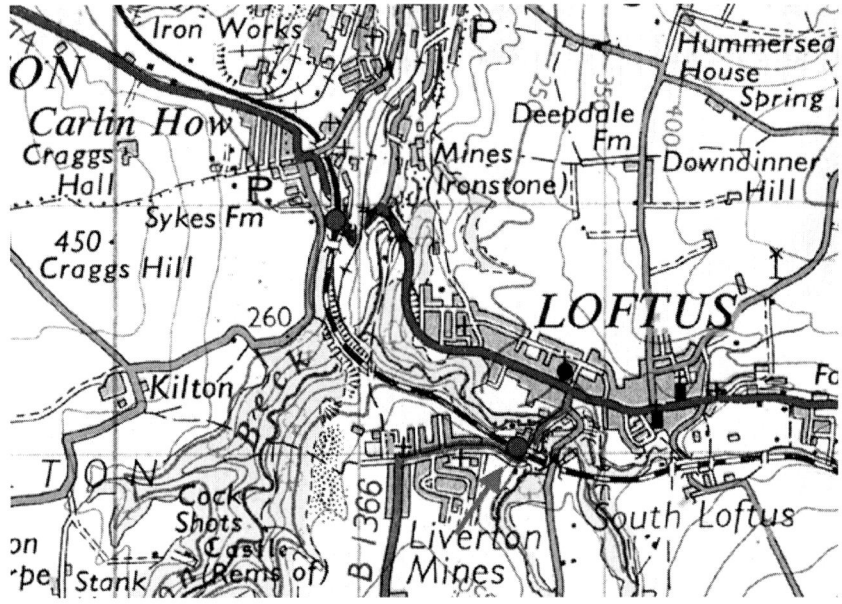

The line in detail from Grinkle [just off the map to the right] to Loftus (also showing Skinningrove station, near Sykes Farm). O. S. map (7th ser., 1955) showing Loftus station.

Loftus station. *J.W. Armstrong/Armstrong Photographic Trust*

Loftus station. *Author's collection*

With 'five on', Riddles class '4MT' No. 80120 of Whitby shed (50G) pauses at Loftus before continuing along the line to Whitby and Scarborough. *D.S. Lowther*

Operating the line

The line was difficult to operate, given the severity of the gradients, the sharp curves and the ever-present sea frets. Speeds were restricted over much of the 16½ miles of the line; even so, Scarborough could be reached far more quickly from Middlesbrough and Stockton than by road. On the branch itself, though, speed was not the most important factor. To all intents and purposes, the line should be viewed as a late-Victorian enterprise, giving the opportunity of travel to local people where none had existed before.

That the line was difficult to build, difficult to maintain and difficult to operate only made it more attractive to those who travelled upon it. Railway 'enthusiasm' as we know it today did not really exist until the 1950s and it was not until its very last years that such enthusiasts began to visit the line. Railway closures, though not as common as they would become six or seven years later, were gathering momentum and the closure of the Whitby – Loftus line was only one in a series of North Yorkshire rural closures which had begun as early as the 1930s. The lines from Gilling to Malton, Scarborough to Pickering via Thornton-le-Dale, Battersby to Picton, Northallerton to Hawes Junction, and Pilmoor to Pickering via Helmsley had already gone. The line's legacy, though, was

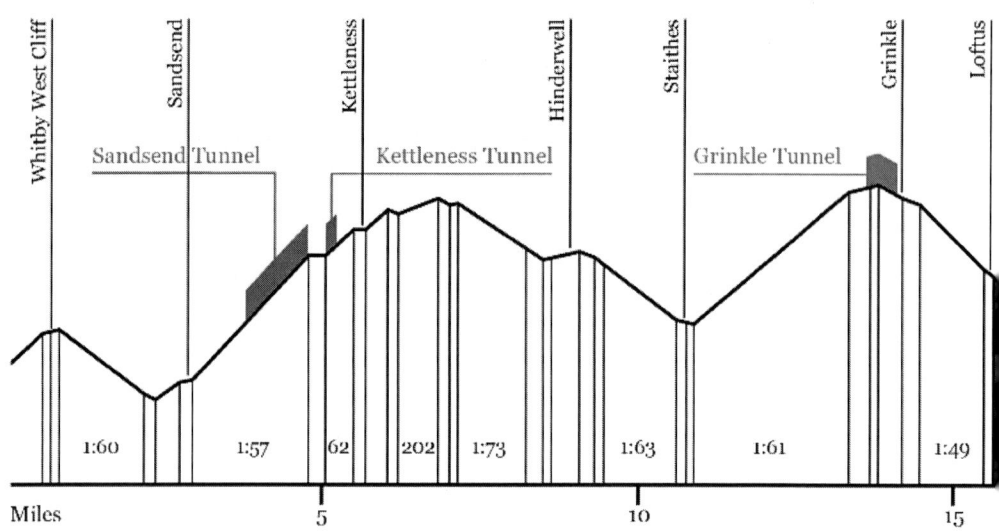

The gradients of the Whitby-Loftus line   *Forgotten Relics of an Enterprising age*

Last day at Loftus. *K.H. Cockerill/Armstrong Photographic Trust*

that it left an indelible impression upon all those who travelled upon it, for who could forget their first journey over the deep valley spanned by Staithes viaduct, the splendid view over the sea as the train entered Kettleness from the north, the breathtaking momentary glimpse of the wild, raw, ancient landscape of the majestic cliffs and roaring waves as the train passed along the 300 yard stretch on the cliff face at Holmsgrove between the Kettleness and Deepgrove tunnels; and who could ever forget the first glimpses of Whitby and Sandsend as the train left the cutting to the south of Deepgrove tunnel and crossed the Steeping Pits embankment? The journey thus became more than just a means to an end, to reach a destination, but instead became part of the very spirit of England.

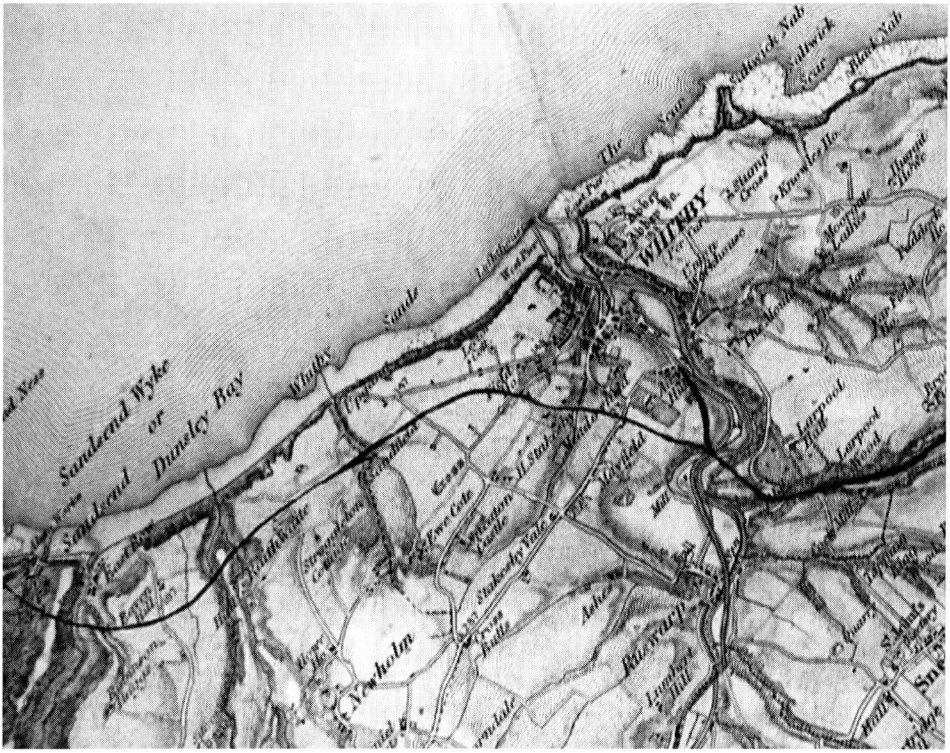

1863 map. Whitby to Sandsend.

1863 map. Sandsend to Staithes.

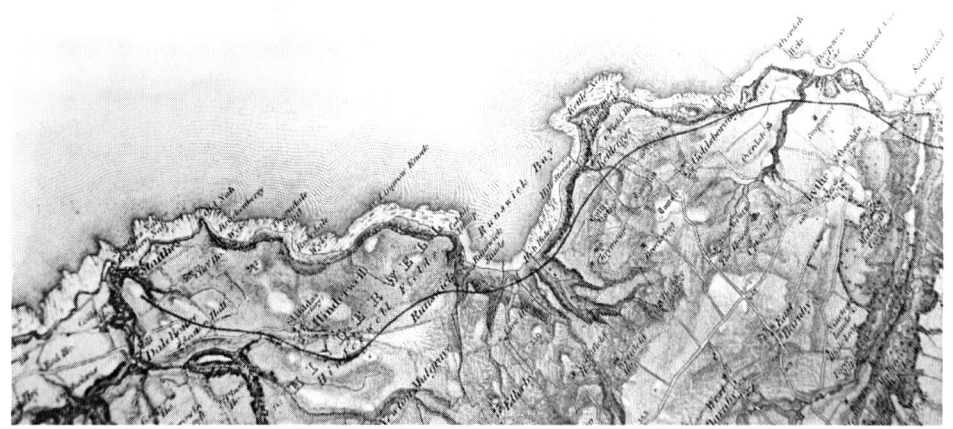

CHAPTER TWO

EARLY MAPS AND PLANS

The earliest plan for a line between Whitby and Loftus was that of the 'Scarborough, Whitby, Stockton-on-Tees and Newcastle and North Junction Railway', which was to run from Scarborough via Whitby and Guisborough to Stockton and thence north. This was one of many wildly impractical schemes which covered the north-east 'with a complicated network of imaginary lines' which emerged during the period of the 'Railway Mania' of the mid-1840s.[1] Nothing came of this idea.

The next proposal for a line along the coast northwards from Whitby was that of the 'Scarborough, Whitby, and Staithes Railway' which, in 1864, was rejected by committee in the House of Commons.[2] On 11th November, 1863, a printed notice was 'hereby given that application is intended to be made to Parliament in the next session, for an Act to incorporate a Company with power to make and maintain the Railways following....'[3] This railway was to begin 'in the township of Falsgrave [Scarborough]....by a junction with the North Eastern Railway at a point on the west side of that Railway 160 yards or thereabouts south-west of the bridge which passes over that Railway at Love Lane'.[4] This is more or less where the line to Whitby would begin when the Scarborough-Whitby branch was opened in 1885. The 1863 plan intended the line to terminate 'in the township and parish of Hinderwell....on the south side of Old Stuble Hill, and 300 yards or thereabouts south of the public house called The First and last Public House, on the public road leading from Hinderwell to Staithes'.[5] There was also to be a connection with the Whitby and Pickering line near Bog Hall. To deal first with the 'main' line: according to the map the line between Whitby and Sandsend would have been built a little further inland than was ultimately the case. Between Sandsend and Kettleness the course was more or less similar to that which was eventually built, although the map gives no indication of any tunnelling or, indeed, any cliff-edge work. From Kettleness to Hinderwell the line would have taken a route close to the sea and Runswick village than was eventually constructed. The route on the plan appears to have been rather carelessly drawn, giving little indication of any engineering difficulties (like tunnels and viaducts) that might arise.

The equally important connection between the Whitby-Pickering line and the proposed Staithes line was interesting, and somewhat different, from the final route when opened in 1883. According to the application, the connection was to begin with 'a junction with the said Whitby and

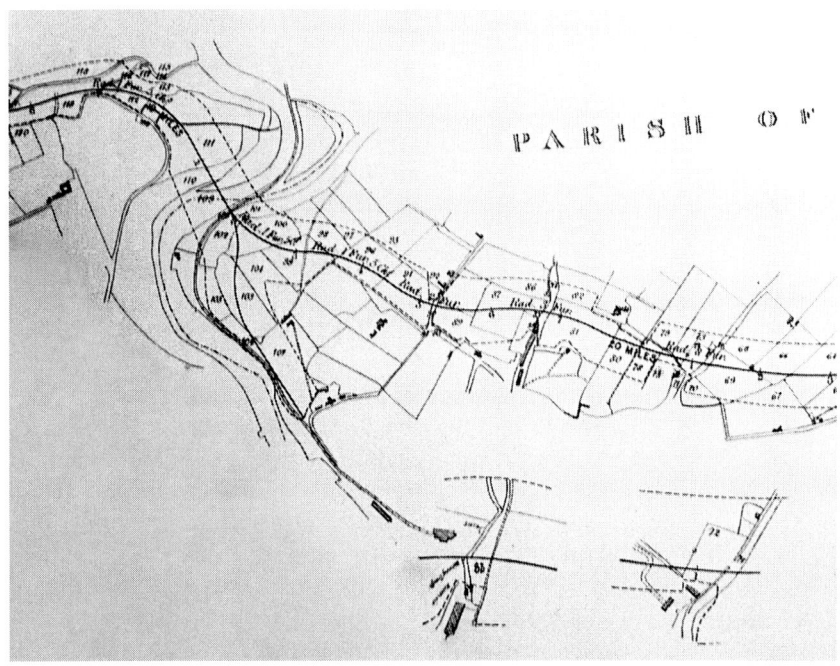

1863 Plan of 'Railway No.2' (the connecting line between the Whitby-Pickering line and the proposed Scarborough-Staithes line). Bog Hall to Larpool.

Pickering line, at a point near to and on the south side of the place where that Railway crosses Waterstead Lane on the level'.[6] Where the 'Railway crosses Waterstead Lane' is more or less where Bog Hall Junction was ultimately constructed. According to the map and plans, this section of the line would have met the 'main' line at a point just west of where – again more or less – Larpool viaduct was finally built (*see page 32*). It is interesting to note that in the 1863 plan there appears to be provision for a high-level bridge or viaduct over the river Esk whereas the 1866 plans clearly indicate a low-level bridge.

It may be said that the history of the line during the time it was in the hands, entirely or partly, of the independent company may be divided into four stages: 1866-71, 1871-5, 1875-83, and 1883-1889. To all intents and purposes the history of the line begins in 1866 with the incorporation of The Whitby, Redcar, and Middlesbrough Union Railway (WRMUR) on 16th July, 1866.[7] The maps for the proposed line are much the same as those for the 1863 venture, with the major exception of the connection between the Whitby-Pickering line and the

WRMUR line. From Kettleness to Loftus the line is in the same position as it was upon completion in 1883, but as in the 1863 maps, there is little indication of any tunnelling or bridging between Sandsend and Kettleness (the map ignoring the topographical problems completely), while the line between (the site of) Whitby West Cliff station and Sandsend ran further inland than the final construction; indeed, the maps show the same alignment as those of 1863. As with the 1863 maps and plans, the most interesting detail is that of the proposed connection with the extant Whitby-Pickering line.

According to the plan the connecting line was to leave the Whitby-Pickering line to the west, just before reaching Ruswarp station, and make connection with the 'main' line which was to ascend at a gradient of 1 in 77 on a steep curve to Prospect Hill: 'A railway commencing....by a junction with the Whitby branch of the North Eastern Railway (NER), about 93 yards north-eastward of a point where that branch railway crosses on the level the road leading from the suspensien [sic] bridge over the River Esk....' Thus any train departing from Whitby (Town) would have to reverse at Ruswarp. This was clearly unsatisfactory.

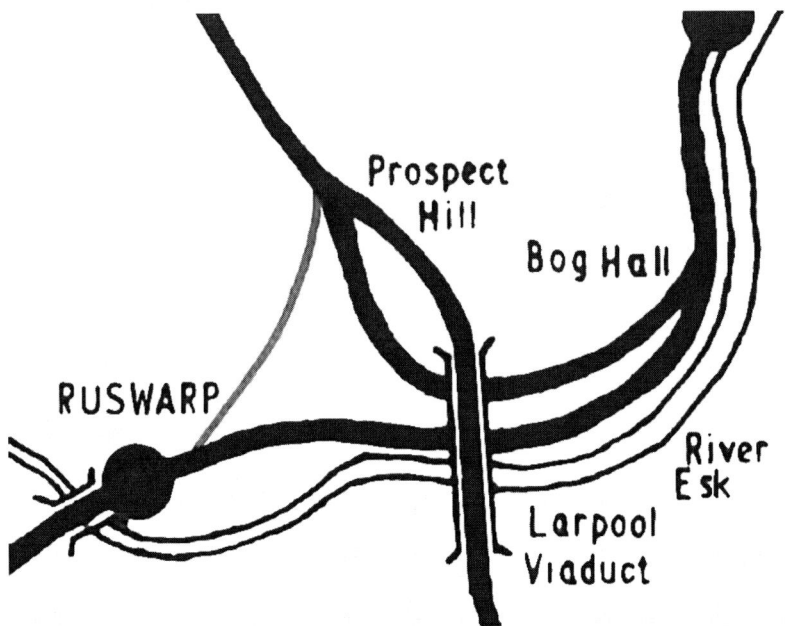

1866 plan. Proposed route from Prospect Hill to Ruswarp. The map also shows the route of the line as finally built from Bog Hall to Prospect Hill Junction. The Scarborough line over Larpool viaduct opened in the summer of 1885.

Rather oddly, it seems to us now, there was to be another line, 'Railway No.2' (*see map below*) which was to be 'all in the parish of Whitby, commencing by a junction with the intended railway hereinbefore described in the township of Ruswarp, at a point in an orchard about 100 yards north of Ruswarp Hall, and thence passing into the township of Hawsket[*sic*]-cum-Stainsacre, and there terminating by a junction with the authorized Scarborough and Whitby Railway at a point near Larpool Wood, about 100 yards north of Crowdy Hall.'

This was a short branch of 3 furlongs and 9.40 chains, descending at 1 in 61, whose purpose was to link up with the proposed Scarborough line, on the far side of the River Esk, which was planned to end at what seems to be the same junction as that of the WRMUR line at Ruswarp. The Whitby-Pickering line would be crossed by a bridge 16 feet high, after which it was planned to cross the Esk on a bridge of four spans 50

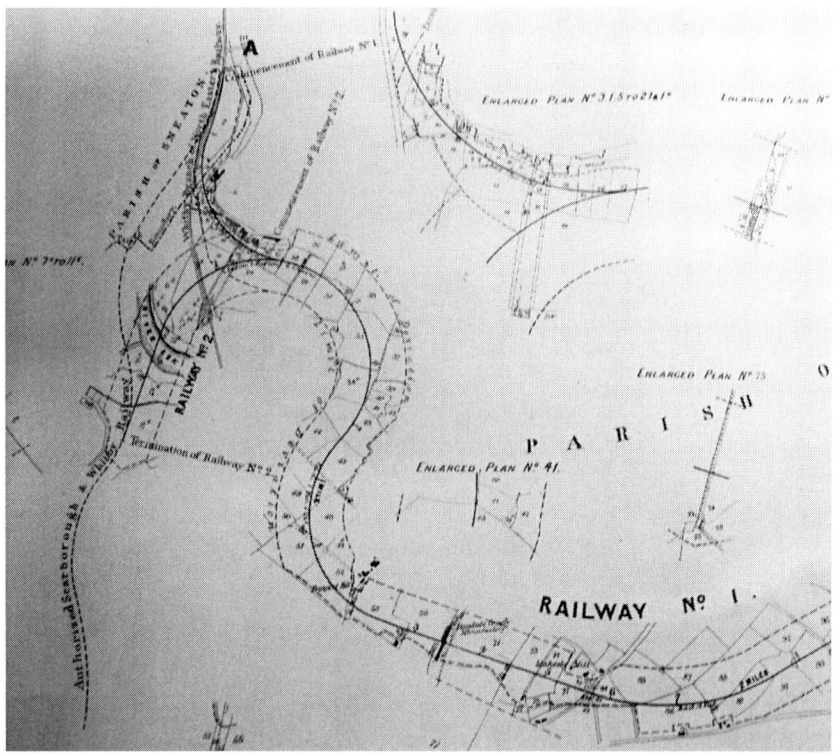

1866 Plan. Proposed junctions at Ruswarp [A] and Larpool. The map also shows the original planned route of the Scarborough-Whitby line which was to join the Whitby-Pickering line at Ruswarp.

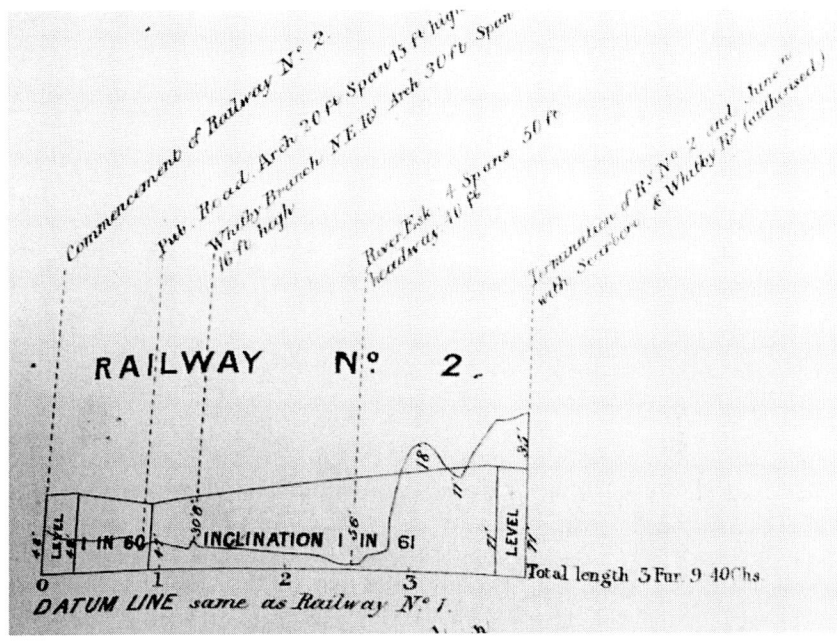

1866 Plan. Proposed bridge over the River Esk.

feet in height. The estimated costs of these lines were, firstly, from 'Ruswarp junction' to Loftus: £211,824. Of this £66,100 was to have been spent on tunnels and viaducts; and, secondly, £13,454 for 'Line No. 2' from Prospect Hill to Larpool.

However, for all the detail of these plans, no actual work was begun on the line until 25th May, 1871, when the first sod was cut at Sandsend. Thus the first phase of the line, that from 1866-71, while laying some practical foundations for the construction of the line nevertheless was never really more than theoretical. Although many problems would confront the builders of the line in the 12 years from 1871 to 1883, it was the necessity of making a sensible connection with the extant NER Whitby-Pickering line and, indeed, to tap the Whitby market for both passenger and freight traffic, that was initially the prime concern of the company, for clearly a reversal at Ruswarp from trains starting from Whitby would be highly unsatisfactory. This problem was overcome during the Parliamentary session of 1872-3.[8] Included in the amended Act was authorization to construct the connection at Bog Hall: 'To make and maintain a deviation railway with all proper stations, approaches, works, and conveniences connected therewith, commencing at a

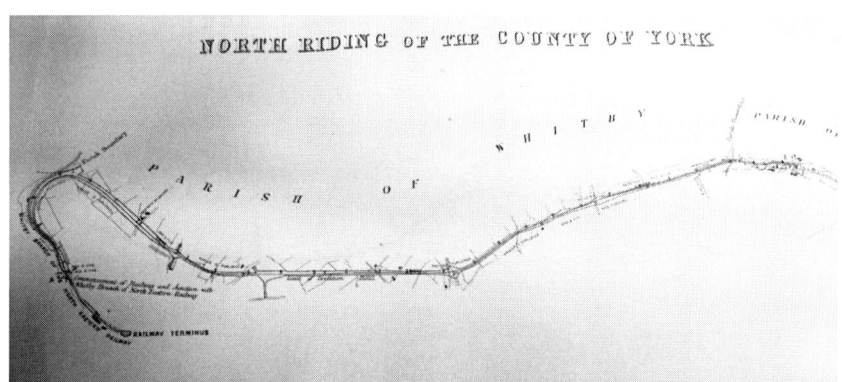

1873 Plan. The route from Bog Hall Junction as finally built.

junction with the Whitby branch of the N.E.R. at a point about 20 yards measured along that branch from the public road level at Boghall....'[9] This was to be the line that was eventually built and which was to remain until the closure of the Scarborough branch in March 1965.

The 1873 Deviation Act also indicates that the line between Upgang and Eastrow was to be much closer to the shoreline (as it was eventually built) and the length between Eastrow and Sandsend was to be lower and close to the shore (again, as it was eventually built).

The line, on the left, from Bog Hall Junction climbs up to Prospect Hill; Whitby Gasworks and the line along the Esk Valley are on the right, August 1957. *Oakwood collection*

CHAPTER THREE

A DIFFICULT YEAR IN THE HISTORY OF THE WHITBY, REDCAR & MIDDLESBROUGH UNION RAILWAY

A note on spelling

In 1873 the small town known today as 'Loftus' was universally spelt 'Lofthouse'. Local accents slowly changed the name to the more demotic 'Loftus' and by the opening of the line in 1883, that is how the town was known - as it has been ever since.

Introduction

The Whitby, Redcar, and Middlesbrough Union Railway Company (WRMUR), throughout its short history (1866-89), was almost always short of funds. The line was extremely difficult and expensive to build and it is remarkable that it ever reached completion. By 1873 problems were mounting. While the viaducts were more or less in place, the proposed tunnelling was seriously behind schedule, with landslips and unforeseen difficulties having a deep impact upon costs. The line, contracted for completion in the midsummer of 1873 was a long way behind schedule; the availability of adequate manpower was always uncertain, and funds, in the form of investments and loans, were becoming scarcer and scarcer. There can be no doubt that the Board of the Directors were subject to severe financial concerns, and penny-pinching had become chronic. This chapter considers in detail, using only the primary sources, the acceleration of the line's problems in the 'long year' of 1873 (they dragged on long into 1874), leading to the agreement with the North Eastern Railway in 1875 to complete the line.

Work begins

On 29th April, 1871, the Board of Directors of the Whitby, Redcar, and Middlesbrough Union Railway Company agreed that a contract with John Dickson be approved and ordered to be sealed.[1] In an atmosphere of optimism and some formality, the ceremony of cutting the first sod of the new railway line was performed on 25th May, 1871 by the Dowager Marchioness of Normanby.

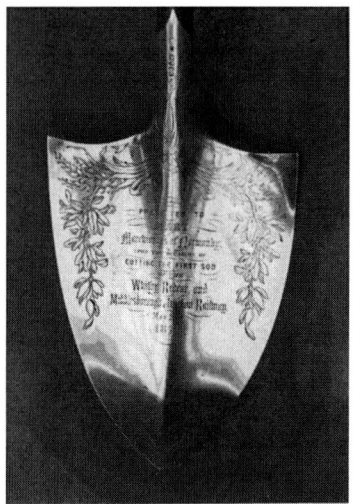

The spade and wheelbarrow used at the formal cutting of the first sod by the Marchioness of Normanby, 25th May, 1871.
By permission Whitby Literary and Philosophical Society, Whitby Museum

However, the Directors, the Engineer and, more importantly, the contractor had underestimated the difficulties of building a railway over such difficult terrain. While the construction of the viaducts continued apace, the problems and, more importantly, the cost of constructing the Easington tunnel and the way around Keldhowe (the largest cliff between Sandsend and Kettleness) began to eat into the company's capital very rapidly. As the months passed, the likelihood of the railway being completed by the Midsummer of 1873, as agreed in the contract with John Dickson, seemed at first unlikely, then impossible.

The contradictory nature of the primary sources

The primary sources for this period are extremely interesting, not only in showing the increasing difficulties experienced by the contractor, Engineer and the company itself, but by the way in which they contradict each other. A key document which goes some way to explaining some of these contradictions is a letter from Arthur Hamand the co-Engineer of the line.[2] The earliest engineer's reports were signed by J. H. Tolmé but, from May 1872 when (as Hamand later wrote 'as I had from [then] the works of the line under my care') these reports were co-signed by both Tolmé and Hamand. Hamand had been stung quite badly by certain remarks made about the lack of progress of the line at the Directors, meeting of 31st March, 1874 and was eager to redeem his somewhat tarnished reputation,

> As to the engineers' reports I wish to mention that every Report issued previously to September of last year (1873) was altered to suit the views of the Board. Those sent by Mr. Tolmé and myself not being sufficiently flowery. In March 1873 the report forwarded by me was eventually altered in character and at the half-yearly meeting when I expressed my extreme surprise and protested strongly against it. It is scarcely fair or generous to refer to these reports as they are referred to in the last half-yearly report after the engineers' reports had been altered as above described.[3]

Indeed, the discrepancies between these reports and the actual state of the line at the beginning of 1874 is one of the more startling aspects of its history. Hamand's comment goes some way to explaining this difference. The statements of the Directors, the sacking of the contractor and the Engineers, however, must be seen in the light of T. E. Harrison's memorandum of 1883.[4] This damning document throws a flood of light upon the construction of the line up to the temporary cessation of works in 1874; indeed, it is remarkable that the North Eastern company ever considered taking over the line at this time. It seems that the booming fortune of other Cleveland lines was the deciding factor.

According to Harrison, almost everything was wrong:

> All the bridges with the exception of one were so defective, and in such a dangerous state, that three were obliged to be taken down and rebuilt before the contract was let, and plans had to be prepared for rebuilding or strengthening the others, and nearly every abutment of the viaducts had to be taken down and rebuilt. The culverts were in many cases inadequate to carry off the water. The slopes of the cuttings and embankments were in no case sufficient, the consequence being that slips had taken place to a great extent, and the big cutting at Whitby [*presumably Prospect Hill*] was filled with slurry to the depth of 12 feet.
>
> This state of things coupled with the fact that there were no plans or sections that could be relied upon necessitated a complete re-survey of the whole line, and cross sections of every cutting and embankment, so far as they had been executed, a labour far greater than would have been required had no works been executed at all. It may here be stated that in some cases the surveys were so inaccurate that had the original tunnels been completed, they were so out of line with each other, that no junction could have been made with them at all.

So important is this memorandum for an understanding of the line's history while under the aegis of Messrs Dickson, Tolmé, and Hamand that it must be read in full. It provides the yardstick by which other primary sources, however contradictory, must be judged. Thus, bearing in mind these contradictions and the finality of the Harrison memorandum a clearer picture of the difficult year of 1873 emerges. The Engineers' reports do, then, offer a 'flowery' picture. For example, the Directors' report of 12th October, 1872, while acknowledging a 'great scarceness of labour owing to the harvest and iron mines in the neighbourhood', nevertheless confidently predicts that 'the works on the whole twelve miles from Whitby to Staithes are altogether so much ahead of the remaining four and a half miles and in such an advanced state that I see no reason to doubt this portion of the line will be ready for opening in the course of next summer [1873]....the remaining four and a half miles will probably take some time longer to complete.'[5]

Six months later the Directors were beginning to paper over the cracks. In his statement to the shareholders on 4th March, 1873 the Chairman found it necessary to dwell upon the 'unfavourable weather' of the past month. The Engineer's report (still signed only by Tolmé) mentioned, almost in passing, that 'the earthworks between Sandsend and Kettleness, which were somewhat behindhand as compared with the rest of the line, have made most progress during the last half-year and they are now well in hand'.[6] There was no mention of this part of the line in the previous report. As far as the Directors were concerned, any problems so far were the result of either shortage of labourers or inclement weather. Then, concerning the meeting of 4th April, 1873 the minutes note the appointment of Mr Robert Hodgson of the North Eastern Railway as Consulting Engineer. Once again we are confronted with contradictory primary sources, as Hodgson's report gives the line a clean bill of health.[7] As Consulting Engineer – and employed by the North Eastern company – Hodgson would not have been under any pressure to produce a 'flowery' report as Hamand claims to have been. Even so, some claims in the letter from Hodgson to the Board of Directors after inspecting the line differ so dramatically from the later pronouncements of T. E. Harrison (Hodgson's superior) that it is difficult to accept them. For example, '....the masonry of all the bridges I examined is very fair average work with which *no* fault can be found.' [*Writer's emphasis*] This directly contradicts T. E. Harrison's 1883 memorandum where he stated baldly that 'All the bridges with the exception of one were so defective, and in such a dangerous state, that three were obliged to be taken down and rebuilt before the contract was let, and plans had to be prepared for rebuilding or strengthening the others....'

Hodgson also wrote that '....from Sandsend to Barnby Dales the work along the face of the cliffs is progressing most satisfactorily. The whole of the line along the cliffs is on solid ground and will turn out a decided success....' Again, this assessment directly contradicts not only Harrison, but the North Eastern company's Directors who, when replying on 23rd October, 1874 to the Whitby company's requests that they (the NER take over the construction of the line) demanded that the course of some of the present line be altered: '....they cannot advise their shareholders to adopt the present line with a view to its completion or become subscribers to it.' However, if the route of the line were to be altered then '....the N.E. Board are prepared to look favourably on the construction of a line of railway properly laid out to meet the requirements of the district'. What this meant was that the line around the cliff should be abandoned (as indeed it was) and that a line be built much further inland (which it was not, the problem being solved by a tunnel slightly further inland from the cliff).

The difficulties emerge into the open

Now, however, things began to change rapidly: a note of disquiet enters the minutes of the Directors' meetings as the year progresses. Those of the meeting on 11th July, 1873 ominously state that 'Much discussion took place on the condition and progress of the works and especially with reference to the iron viaducts and to the backward state of the works at the tunnel....'[8] On 17th July, 1873 the ominous note continues: 'Much discussion took place with reference to the condition of the works and as to the means of hastening the completion of the same'.[9] On 25th-26th July it was clear that not only was Dickson's work being viewed with concern, the Engineer was also being scrutinized for, the minutes noted 'After an inspection by the Directors on the condition of the works it was resolved that Mr Hamand be instructed forthwith to furnish Mr Hodgson with such plans, sections, and particulars of work already executed as will enable him to check Mr Hamand's calculations both of the work completed and yet remaining to be done'.[10]

Then the Directors' Report of 18th September, 1873 (a much more 'for public consumption' document than the minutes of each Board meeting and thus more circumspect) made clear the true state of affairs: 'The Directors regret that the progress made during the last six months has not been more satisfactory....the time limited for the completion of the Contract has already expired, and after making every allowance for the difficulties referred to in the Engineers' Report, they are not satisfied that the Contractor is free from blame'. Labour shortages, difficulties with supply of materials and problems with the Easington tunnel where a slip had caused 'a great deal of trouble and delay' were given by the Engineers as reasons for the tardiness of the works but '....notwithstanding the difficulties encountered by the Contractor, it must be admitted he has scarcely made so much progress with the whole work as could be desired.'[11] It was probably at this meeting that the disquiet which had been barely beneath the surface came into the open. The catalyst seems to have been the contractor's request for a further £2,000 over and above that certified to him by the most recent Engineer's certificate. In return for this advance the Directors demanded security, which was offered to them in the form of two locomotives.[12] Dickson's sacking was now imminent for on 21st November, 1873 the Directors accepted '....a certificate signed by Mr Tolmé and Mr Hamand that the Contractor had failed to proceed according to his obligations under the preliminary contract made the 3rd day of May 1871 between the Company and the Contractor.' Further to this, Dickson was asked 'to show cause why the Company should not take the further works out of his hands and take and use the plant and materials for the purpose of finishing the works themselves or employing another contractor to complete them.' Dickson was then advanced £1,000 to 'prevent the stoppage of the works.'[13] It was this concern – that the works continue – that led to the arrival in Whitby of George Fraser, the son of the Company Secretary, James Fraser.

The 'My Dear Father' letters

With the sale of the completed and running railway to the North Eastern company in 1889, most of the documents held by the Whitby company came into the possession of the North Eastern. Among these documents are 12 letters sent from The Royal Hotel, Whitby by George Fraser to his father. James Fraser was the Secretary of the Whitby, Redcar, and Middlesbrough Union Railway, having been appointed on 3rd April, 1871, with a salary of £150 p.a. The family home was at 3, Middleton Road, Camden Road, London N. James Fraser's office was at 9, King's Arms Yard, London, E.C. These letters are exceptionally important in that they give a clear idea of day-to-

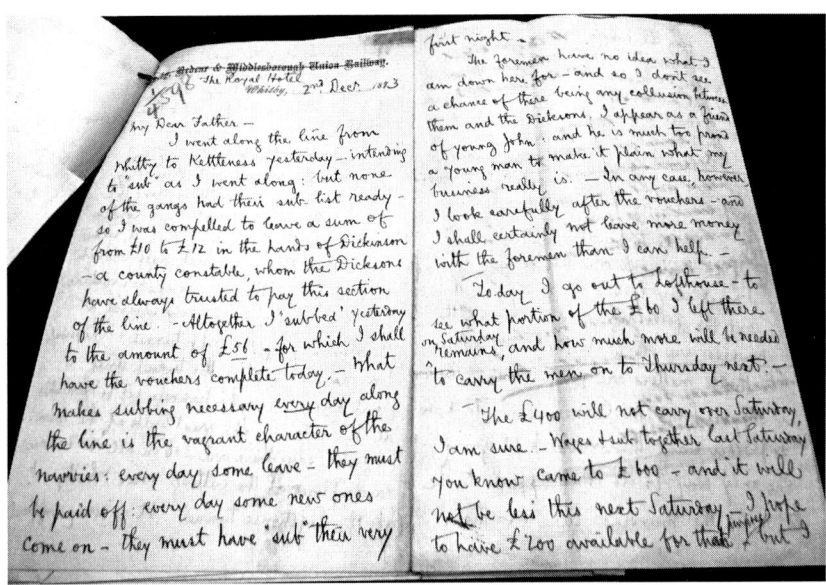

TNA RAIL 743/9. Part of the letter of 2nd December, 1873 from George Fraser to his father James Fraser, the Secretary of the Whitby, Redcar, and Middlesbrough Union Railway.

day life in the construction of the line, the view from the 'cutting/embankment/tunnel face' as it were, and show the difficulties and frustrations of trying to hold together such a large and disparate work force. The letters span only a fortnight in the history of the line, but they are written at a time of great uncertainty, upheaval, and change. The Directors were determined to get rid of Dickson while at the same time ensuring that the day-to-day construction continued. The Secretary, John Fraser, sent his son to Whitby to ensure continuity of pay for the workmen during the upheaval caused by Dickson's sacking (which was confirmed on 6th December).

As well as providing a sharp insight into the difficulties in constructing a railway in the late 19th century, the letters often quite unconsciously throw light on conditions of life for both working and middle-class men 140 years ago. It may be argued – and the present writer would certainly do so – that the fundamental troubles encountered on the lengthy construction of this line were caused not only by both an inadequate contractor and engineer, but by the penny-pinching policy of the company itself. Above all, these letters show how quickly considerable sums of money could be spent in such a large construction project. It appears from the letters that the Frasers were well aware of the imminent departure of Dickson and there is an undertone of subterfuge running through them (in that the Dicksons had not been informed and that it was necessary not to cause doubt among the working men, called 'navvies' by George Fraser). Also apparent is a growing friction between Arthur Hamand (the Engineer) and the Dicksons. Neither Hamand nor Dickson appear 'in person' as it were in the letters, they are represented by underlings. In Hamand's case, Messrs Eglinton and Greenwood, and in Dickson's case, his sons. One also gains the strong impression that the longer George Fraser remained in Whitby, the more sympathetic he grew toward the Dicksons and the more confident he became in dealing with the problems he encountered on a daily basis. Mentioned frequently in the letters is the one Director who lived locally, Mr Edward Corner of Esk Hall, Sleights (which still exists). Mr Corner was a Director for most of the period between 1871 and 1889 and in later years became its Chairman. He is certainly the major moving force in the line's construction after these initial problems.

We do not know how long George Fraser spent in Whitby; neither do we know anything about him, although internal evidence suggests that he was a young man, quite possibly in his early twenties, still living at home, though with wider ambitions than working for his father. Touchingly, and no doubt conforming to the middle-class family mores of the time, he gives the impression of a dutiful, obedient and loving son, not afraid to ask his father's opinion when in doubt over an issue. At the time, his principal occupation may very well have been as a junior in his father's office, perhaps learning the trade and being groomed to take over affairs after his father's retirement; but this perhaps is a speculation too far. Arriving in Whitby in November 1873 (an area of the country to which he had not previously been) he was immediately attracted, 'I need hardly tell you that I am delighted with the beauty of this north countree [sic] and enjoyed my drive to Lofthouse very much.'[14]

George Fraser realized quite shortly that ensuring the continuity of pay and thus keeping the 'navvies' at work was to be a far more difficult and complicated task than he at first realized, for he was confronted with something quite out of his experience: the 'sub' system. Simply, a 'sub' was an advance on the week's wage, but many of the navvies on the line used the system so regularly that they were, in effect, being paid by the day. George Fraser did not approve, but there was nothing he could do. 'This "sub" system is a shocking one but the Dicksons say it is the only way by which the men can be kept together.'¹⁵ In a second letter the same day, George Fraser (clearly taken aback by the situation) wrote, 'this "sub" system is a most pernicious one, I think. The men necessarily lead a hand-to-mouth existence - and the master has to provide a daily pay.'¹⁶ His comments on this unfortunate system also throw considerable light on the navvy lifestyle, 'what makes subbing necessary every day along the line is the vagrant character of the navvies: every day some leave – they must be paid off: every day new ones come on – they must have their "sub" the very first night.' [*writer's punctuation*].¹⁷ So demanding were the working men for their 'subs' that most of George Fraser's letters are taken up with constant demands for more money from his father; when that was not immediately available he had to visit Edward Corner at Sleights (sometimes missing him there and catching up with him in Whitby) for emergency loans. Indeed, subbing appeared to be the only way to ensure the works kept going and, wrote George Fraser, 'The only way to do that appears to be to make the navvies sure of their "sub" from day to day. They are most improvident fellows and, as I told you, they regularly "sub" some of them within a few pence of their complete pay'.¹⁸ Nevertheless because of these incessant financial demands – and the need to provide evidence for them – there is considerable information concerning the numbers of people at work on the line. There was a discrepancy, or, at least, a misunderstanding in the numbers being paid on 2nd December. On the one hand George Fraser informs his father that 'the great bulk of the men (Eglinton says there are 516 in all and 20 horses)'¹⁹, while in an undated letter (there is a page missing but it is likely to be 3rd December) Fraser jnr tells his father that 'Greenwood and I actually paid 625 men and boys.'²⁰ 'Whereabouts on the line were the men working? Only a few were at Keldhowe.'²¹ The majority of the workforce was 'employed in the cuttings near Whitby (almost certainly at Prospect Hill) and at the Lofthouse (i.e. Easington) tunnel.'²²

On 5th December, 1873 George Fraser in his now daily letter to his father related that, on that day, he 'was out at the Lofthouse end yesterday and saw the whole of the works in progress from the Easington (i.e. Lofthouse) tunnel to the point where the junction with the North Eastern (i.e. an end-to-end connection at Loftus) is to be put in. In the three cuttings somewhat over 100 men and boys were employed with trucks and horses and altogether the progress seemed to be at a very satisfactory rate.'²³ George Fraser's days were, in the main, taken up with visiting the line and paying wages and subs.

However, he found it impossible to visit the entire line in a day and he and Greenwood divided up the task, Fraser taking the Whitby to Ellerby section, while Greenwood dealt with the Ellerby-Loftus section. However, it should not be thought that the workforce was mainly comprized of navvies. Fraser jnr wrote to his father that '....looking to the nature of the works between Whitby and Hinderwell it is quite clear that as many *skilled* workers are needed for this section of the line as for the other. Masons, miners, platelayers, carpenters, engine-drivers and smiths are the best paid workmen and they are numerous enough here....'[24]

Less appealing is the somewhat clandestine element in the letters. It is clear that one of Fraser senior's instructions to his son while in Whitby was not to give any indication to John Dickson (or his sons, or the workmen for that matter) that his sacking was imminent and also to give no indication that the plant and materials were about to be 'taken over' by the company. Dickson, of course, knew that his position was precarious.[25] Arthur Hamand, the Engineer, was also involved in this deception. He had sent Eglinton to the area to 'brand the plant' i.e. to attach evidence of ownership by the company of certain materials on the various construction sites. This somewhat mean-minded policy of the company threatened to backfire when James Dickson (one of the contractor's sons) complained about the branding of some trucks which were his own property. What was worse, Eglinton 'seems hardly to act with the discretion which is to be wished for and it is chiefly his conduct which has set all the navvies, and indeed the town in general, to suspect that something is wrong with the Dicksons. I told him today that he must be very careful not to brand anything which he was not quite sure of being the property of Mr Dickson snr.'[26] This tactlessness on the part of Eglinton had two unfortunate consequences. Firstly, it aroused the suspicion of the navvies to the point where, George Fraser wrote, 'what makes the subbing so heavy is the fact that the navvies are extremely suspicious that something has gone wrong, and so they are determined to get every shilling they can out of the contractor at once.'[27] Secondly, it had the effect of alerting local tradesmen to the difficulties being experienced,'....the tradesmen here will supply the Dicksons with nothing more until past accounts are paid....'[28]

The situation then reached its *dénouement* with the decisions made at the Directors' meeting of 6th December, 1873 and the events along the line which seemed to have occurred that night, after the news of the meeting had reached Whitby. At that meeting, among other items on the agenda, it was resolved 'that the Company do now take the further execution of the works out of the hands of Mr Dickson and take and use all the plant and materials provided by him for the purpose of completing the works also'.[29] A scene of almost wild-west proportions then followed. According to George Fraser 'During the night Hamand's men, Greenwood and Eglinton, have made a raid on the stables and seized all the horses.' Fraser jnr liked the Dicksons: 'The Dicksons have treated me throughout this business in the most honourable manner. I feel quite vexed that they should have been treated thus.'[30] Arthur Hamand's reaction, as represented by Messrs Greenwood and Eglinton did not go down well. George Fraser's next letter indicates that there was, shortly after this, a meeting at York. There is no evidence for a meeting at York at this time in any of the primary sources, so it may very well have been a non-official meeting; nevertheless Messrs Corner and Leeman (both Directors) were there. 'Neither Mr Leeman nor Mr Corner were satisfied with Hamand's behaviour. He seems to have made a point of attacking me....'[31]

After a letter from Whitby, dated 15th December, 1873 the letters stop, either because George Fraser's work in Whitby was done, or because no more have survived. Unfortunately, none of the letters from father to son exist. The letters are windows into a society long vanished and gone, whose concerns nevertheless are instantly recognisable to modern readers, and whose expression of filial love, duty, and respect command an equal respect from the reader. 'But here I think I must close my note for the present. *It's just about half past three on Wednesday morning.* With much love to all at home, believe me, dear father, your affectionate son, George.'[32] [*My italics*] They also give an interesting comparison of life before and after the coming of a railway. On the previous day (a Saturday) George Fraser along with the contractor and his cashier (a Mr Williams) 'drove out in a dog-cart to the Lofthouse and of the works (16 miles or so)....it was nearly ten 'o clock before we got back to Whitby (they had set off at one o'clock)....It was a long drive (nearly three hours each way) which made us so long.'[33]

Once the line had been built and was in operation, the trip from Whitby to Loftus took approximately 50 minutes, according to the 1887 timetable.[34] Today the journey by bus takes 40 minutes. Once a railway was built then very long distances indeed could be covered in a day. Whitby was approximately 250 miles from London by rail in 1873 (changes having to be made at (probably) Malton and (certainly) York). George Fraser was very interested in a post in London at the time and, writing on Friday, 5th December said 'I could easily arrange to be in London for the Mansion House Committee on Wednesday *and return here that same evening*'.[35] Presumably he would have travelled down the day before, but not necessarily so.

The letters give also an insight into the difficulties faced by the navvies. George Fraser was somewhat contemptuous of their lifestyle and improvidence, but he recognized that their lot was not easy, especially when he realized that 'it is very difficult or even impossible for the men to find lodgings in the country which lies between here and Lofthouse, and though Mr Dickson has built huts for them those huts are now quite deserted.[36]

Troubles and difficulties continue into 1874

So Dickson was sacked, and it would not be long (before the end of 1874) that the same fate would befall Tolmé and Hamand (before the end of the next year). Hamand was becoming fed up of the whole business and his frustration came to the fore in May 1874. To return to Hamand's letter to the Directors mentioned earlier in this chapter.[37] Like the Harrison memorandum this letter looks at the rather disastrous situation from another angle. While the Directors were keen to throw all the blame upon the contractor and the Engineers, Hamand was quite clear where the blame lay. The Directors' pennypinching and self-delusion was, in Hamand's view, just as much to blame for the troubles on the line as the failings of the contractor. However, they (the Directors) were not willing to accept any blame and, indeed, sought to implicate the Engineers, '....when after the stoppage of the contractor the cost of completing the line was found to exceed anticipation the Directors without giving any weight to circumstances and facts well within their knowledge immediately laid the whole blame on their engineers so that whatever confidence existed very soon evaporated.'[38] As far as penny-pinching was concerned, Hamand commented that '....during 1873 I gave a great deal of personal attention to the line and many thousand cube [*sic*] yards of earthwork.... the actual price paid to Mr Dickson which up to the end of 1872 only averages about 1/- per cube yard for earthwork certified....the first Schedule allowed 1/6 per yard....no contractor would undertake the earthwork on the whole line now under 1/6 per cube yard. I have selected earthwork as being the largest item in the cost of the line but the same reasoning will apply to other things'.[39]

As always, it was the decision to build the line around the cliffs between Sandsend and Kettleness that was at the root of the problem. Moreover, the method of payment to Dickson was plainly unsatisfactory,

> ....the Directors seem to be unaware of the impossibility of measuring the cliffs and such works and ignore the fact that while we were reducing the prices paid other companies were increasing theirs from 15% to 30% while in some cases contractors were released from their obligations altogether while our contractor was paid partly in stock which he could only realize at a loss. Moreover a contract with a man of little or no capital does not render the duties of engineers those of the easiest in the face of rising prices.[40]

Finally, and Hamand was proved correct in this instance, there was simply not enough money left to complete the works,

> ....in my letter addressed (to) the Secretary dated 21 Aug 1873 I said Mr Hodgson would certainly estimate the value of the work remaining to be done at from £30,000 to £35,000 more than the money we had left to do it with....at the same date in a letter to Mr Leeman I said the line would cost double the available balance if anyone but Mr Dickson finished the line.[41]

The last sentence is very revealing and confirms everything that we now know about the construction of the line under Dickson as described in the Harrison memorandum of 1883. As for the Board's self-delusion, '....I have never denied that more capital to some extent would eventually be necessary and if the Board has been misled it is by their own natural hopes and desires than by any information from me.'[42]

Thus the terrible year of 1873 ended. One might call it a 'long year' as it dragged on well into 1874. To all intents and purposes, and for all the efforts of George Fraser, work stopped on the line for a period of over four years. Work on the viaducts continued but, the Directors reported, '....pending negotiations [sic] with the North Eastern Railway Company the execution of the earthworks and [Easington] tunnel has not been further carried on.' Indeed, it had been 'found necessary to stop the further progress of the works in April last.'[43] However, even the Directors could not hide from their shareholders the fact that they had run out of money, and that this was the main cause of the cessation of works in April.[44] The previous half-yearly Directors' report said as much: 'It was only within the past month [March, 1874] and since the receipt of the accompanying report of the Engineers that the Directors have been informed by them that a considerable increase in capital will be required.'[45] It was this statement which caused the Directors to panic, '….the Directors immediately on being informed of the necessity for considerably increased capital did not feel warranted in expending more of the already authorized capital….the Directors are thus unexpectedly placed in a position of considerable difficulty as to the course they should pursue.'[46] The Directors then told their shareholders that they would shortly submit a plan for the raising of more capital so that the works could continue. This was a fond hope. Only the North Eastern could save the line now.

Details of the Sale of horses belonging to the Whitby, Redcar and Middlesbrough Union Railway Company on 15th May, 1874.

Horse	Sale price	Buyer
Black mare. *Sally*. 10 years.	£35	Mr Stubbs, Skelton.
Bay horse. *Tom*. 9 years.	£22	Mr B. Gardiner, Whitby.
Dapple brown horse. *Dodger*.	£40 10s.	Mr Frank, Pickering.
Brown mare. *Kate*. 8 years.	£40	Mr J. C. Carr
Black mare. *Bonny*. 9 years.	£32	Mr Stubbs, Skelton.
Dapple brown mare. *Dutch*. 10 years	£37	Mr Walker
Black horse. *Charley*. 9 years.	£24	Mr Ralph Bindell.
Black horse. *Punch*. 10 years.	£29	Mr George Mitchell.
Dark grey horse. *Smart*.	£60	Mr Thomas Dickenson.
Chestnut mare. *Gess*.	£20 10s.	Mr Moss Eaton
Bay horse. *Captain*.	£11	Mr Peter Campion.
Bay mare. *Jenny*.	£11	Mr B. Gardiner, Whitby.
Grey horse. *Dick*.	£31	Mr Thomas Newton.
Grey mare. *Beauty*.	£34	Mr Johnson.
Bay horse. *Jack*.	£17	Mr J. C. Carr
Chestnut horse. *Charley*.	£47	Mr Stubbs, Skelton

CHAPTER FOUR

THE VIADUCTS AND TUNNELS OF THE WHITBY- LOFTUS LINE

The Whitby to Loftus line, opened on 3rd December, 1883, was a little more than 16 miles in length, from Bog Hall Junction (26 chains from Whitby station) to an end-on connection with the North Eastern Railway at Loftus.

The first proposal, in 1863, for a line along the coast northwards from Whitby was that of the Scarborough, Whitby & Staithes Railway. According to the map the line between Whitby and Sandsend would have been built a little further inland than was ultimately the case. Between Sandsend and Kettleness the course was more or less similar to that which was eventually built, although the map and the plans give no indication of any tunnelling or, indeed, any cliff-edge work. From Kettleness to Hinderwell the line would have taken a route close to the sea and Runswick village than was ultimately constructed. The route on the plan appears to have been rather carelessly drawn, giving little indication of any engineering difficulties (like tunnels

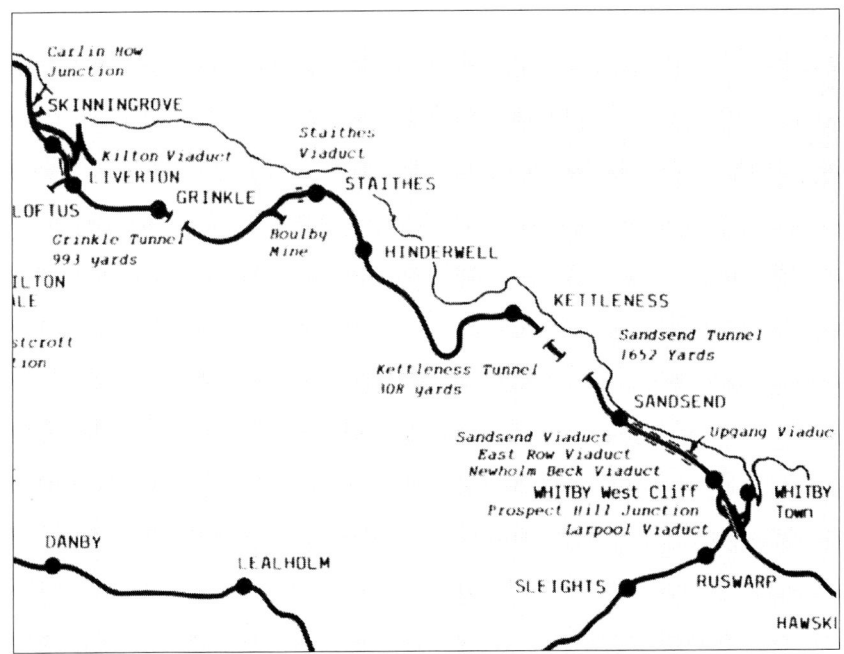

The Whitby – Loftus line showing viaducts and tunnels.

and viaducts) that might arise.[1] It was rejected in 1864 by committee in the House of Commons. The Whitby, Redcar & Middlesbrough Union Railway was incorporated on 16th July, 1866.[2] The maps for the proposed line are much the same as those for the abortive venture in 1864. From Kettleness to Loftus the line is in the same position as it was upon completion in 1883, but (as in the 1864 maps), there is little indication of any tunnelling or bridging between Sandsend and Kettleness (the map ignoring the topographical problems completely), while the line between (the future site of) Whitby West Cliff station and Sandsend ran further inland than the final construction. The estimate for the proposed line gives the total cost as being £235,278 (a vast underestimation as it later turned out) with £66,100 being earmarked for 'tunnel and viaducts'. It is unfortunate that the two were not separated in the estimate but it is known from later costs that tunnelling was far more expensive than the overall costs of the viaducts. The line in its entirety cost £655,077 at £40,492 per mile and was very difficult to construct. Although Parliamentary approval for construction of the line was granted in 1866, the first sod was not cut until 25th May, 1871, with John Dickson as main contractor. In all likelihood this hiatus was caused primarily by the difficulties the company had in raising sufficient funds. However, the terrain upon which the line had to be built included ravines of post-glacial streams; the valley sides of which were steep and often unsuited to railway construction except with the use of heavy and expensive engineering works.[3] This terrain, and the difficulties involved in bridging it, are clearly shown in the photographs of the viaducts, while operating costs were heavy, owing to the nature of the terrain and the continuing maintenance needed on the three tunnels and five viaducts along the short 16 mile line.[4]

The viaducts

The first of the viaducts in the direction of Loftus from Whitby was at Upgang, about two miles from the commencement of the line at Bog Hall Junction. While there are many photographs of the other four viaducts on the line, those showing Upgang viaduct are hard to come by.[5]

THE VIADUCTS AND TUNNELS OF THE WHITBY-LOFTUS LINE   55

Last train to Whitby on Upgang viaduct.   *David R. Smith*

A little further along the line was Newholm viaduct.   *Author's collection*

Within half a mile of Newholm was Eastrow viaduct.   *N. Cholmondeley collection*

Shortly afterwards the trains passed over Sandsend viaduct.   *J.S. Gilks*

THE VIADUCTS AND TUNNELS OF THE WHITBY- LOFTUS LINE

Finally, the *pièce de résistance*: the Staithes viaduct.
J.W. Armstrong/Armstrong Photographic Trust

The design and construction of the viaducts

The viaducts were manufactured at the Skerne Ironworks, Darlington which enjoyed a chequered career: opened in 1864, it closed in 1875 and re-opened twice, in 1876-79 and 1880-82. The designer of the viaducts was John Dixon (not to be confused, as has happened, with the main contractor John Dickson).[6] The contracts (or 'articles of agreement') were made on 1st December, 1871 and 25th January, 1873. The price for the erection of the five viaducts along the line was £23,452 (well over one million pounds in today's money). *The Engineer* of 14th March, 1873 reported 'We have noted with considerable interest the series of viaducts on the new Whitby, Redcar and Middlesbrough Union Railway, constructed for the company by Mr John Dixon from his own designs but under the superintendence of Mr J. H. Tolmé, the engineer.'[7]

Fortunately, evidence is plentiful for the history of the railway and its viaducts between the years 1871-1889.[8] There are three key sources for the early history of the viaducts with special reference to their cost, the dates of their erection on the line, and the problems they caused before the line could be opened.[9] Arguably the central text is that of the Harrison memorandum which was deeply critical of their design and construction (*see Appendix One*). However, this memorandum was written in 1883, and served to justify the late opening of the line by the

NER. The very few historians who have consulted the primary sources accept it at face value; the present historian on the whole agrees with that acceptance.

The viaducts were in place very early in the line's construction. Unfortunately the contract between Dickson and Dixon has not survived; however, it would have been sensible, and not unusual, to bind Dixon to early completion in order to allow Dickson access along the line. The reports of the Engineer, Tolmé, while not always reliable, provide what seems to be a fairly accurate timeline. His report to the Directors of the WRMUR company on 9th May, 1872 (less than a year after the first sod had been cut) informs them that (between Whitby and Sandsend) 'The only heavy work on the section consists of three viaducts, the masonry for which is fast approaching completion *and the iron superstructure is now in course of delivery.*' On 9th September, 1872 Tolmé reported 'One iron viaduct on this length (the first seven miles from Whitby) is finished, and two other iron viaducts and two short timber ones are in the course of construction'. The timber viaducts mentioned are of interest for they were to be built on what soon became the abandoned stretch of line around the cliffs between Sandsend and Kettleness.[10]

The next five months were productive for, on 21st February, 1873 *The Engineer* reported 'The whole of the iron viaducts are now completed, with the exception of the one at Staithes, which is, however, in course of erection.' That this was the case is shown by a most important source, unused by any other historian of the line, which not only gives technical details of the viaducts themselves but includes a vivid and dramatic illustration of Upgang viaduct which, at the time of the publication of the article, can only just have been completed.[11] However, clouds on the horizon were beginning to darken. The Directors' minutes for 11th July, 1873 record that 'much discussion took place on the condition and progress of the works and especially with reference to the iron viaducts and the tunnel.'[12]

The building of the Upgang viaduct was successful for, in his next report of 1st September, 1873, the Engineer, J. H. Tolmé reported that 'One of the piers of the Upgang viaduct, over which trains have been running some time, has been recently weighted with double the maximum working load (viz. 180 tons) and had stood the test in the most satisfactory manner, there being no signs of defect or failure in the slightest degree.'[13] Things were not going entirely smoothly elsewhere for, reported Mr Tolmé on the same date 'The work at Staithes viaduct has been almost suspended the last few months pending some slight modifications in the designs of the larger spans. These are now settled, and the work will be proceeded with'.[14] Unfortunately, time was running out for Tolmé and Dickson for, according to the original contract, the line was to be completed within two years 'by midsummer 1873' and, by September 1873, that time had passed. In December 1873 Dickson was sacked, and by late 1874 the same fate had overtaken Tolmé.

There are two likely reasons for this state of affairs: firstly, the poor quality of work produced by Dickson of which the Harrison memorandum provides the most compelling evidence and, secondly, the impoverished state of the company's finances. That these finances were in a very distressed state is made clear in a very important document: a long letter dated 5th May, 1874 from Mr Arthur Hamand who had, from May 1872, become joint Engineer with J. H. Tolmé. Hamand was clearly unhappy with the company and, indeed, while co-signatory with Tolmé to the half-yearly Engineers' Report to the shareholders of 31st March, 1874 does not appear on such a document again, indicating his resignation or dismissal (neither of which are mentioned in the minutes of the Directors' meetings). In this letter he is severely critical of the company's financial situation, making clear that the company's penny-pinching has considerably exacerbated the problems in completing the line.[15]

The sacking of Dickson caused problems for John Dixon, especially as his contracts for the building of the viaducts had been made with the former. Dixon wrote to the Board of the Whitby company on 9th January, 1874 stating that 'I am now prepared to complete the viaducts according to my agreement with your late contractor John Dickson for the sum of £3,152, being the balance as shown by you, due to me.' He went on to say that the real balance was actually £5,252 'a portion of which has been retained by you in the shape of retentions according to your agreement with him....'[16] These figures indicate that, by early 1874 almost three-quarters of the viaduct work had been completed. The Board seemed to acquiesce with Dixon's demands; the minutes of the Directors' meeting of 9th February, 1874 note that 'The correspondence and arrangements made with Mr Dixon for the completion of the iron bridges were read and confirmed.'[17]

In order for Dixon to be paid, the Engineer (Tolmé) had to issue a certificate which had the effect, among other things, of authorizing payments to contractors. By March 1874 Dixon was becoming impatient for payment and on the 26th wrote to the Board asking for £1,500. He complained that he had had no certificate since the previous July

'....why the Engineer should not issue one regularly every month I cannot understand as it is stipulated for in the contract....the work has been going on continuously'. Nothing happened. John Dixon finally lost patience and on 8th May, 1874 sent a very tart letter indeed to the Board, demanding that 'your engineer be instructed to do his duty'. It was now nearly a year since he (Dixon) had received a certificate although 'I have steadily kept on with the works and I think I am entitled to have consideration'. It is apparent, then, from these letters that although the rest of the work on the line had ceased, work on the viaducts was still continuing.[18] This is confirmed by Tolmé's last report in that he wrote '....the viaducts on the line have been steadily progressing and are now nearly completed.'[19] What is also apparent is that the Board of the company was in some disarray, caused most likely by the lack of funds.

By now negotiations with the NER to take over the construction of the line were pending. However, it was not until a year later, after the agreement with the NER, that Dixon's demands began to be met. The minutes of the Directors' meeting of 20th July, 1875 confirm that '....a cheque for £2,000 be paid him', while the minutes of the next meeting, held on 27th August, 1875 include the resolution that'....a further £1,000 on account of his contract sum of £3,152 be authorized'. Even so, Dixon was not satisfied. The matter dragged on for a further three years until, in September 1878, the minutes of the Directors' meeting of the 13th declare that'....Mr Dixon's claim had been settled for £4,500 of the Company's stock transferred to him'.[20] By 1889, however, this stock had become almost worthless. Nevertheless the viaducts were completed and, to all intents and purposes, John Dixon had (ultimately) been paid.

Before moving to a discussion of the problems presented by the viaducts in the early 1880s, the cost of their construction must be considered. The main source for their cost is the Engineer's Reports of 1872 and 1873. These reports give (usually at monthly intervals) the amount of money spent on various aspects of the line. The costs seem to be cumulative (annually). Thus for 1872 the amount spent on the viaducts was £5,650, while in 1873 the amount spent was £7,500. In today's money the amount spent on the viaducts in these years is approximately £600,000. These figures are important, for they partly explain why a deviation proposed in 1875 was not taken up. Briefly, this plan was to abandon the section between Sandsend and Kettleness and to construct a deviation inland from Whitby (West Cliff) to Hinderwell. It is possible that it came to nothing because the viaducts would have been rendered redundant and the money spent upon them wasted.[21]

THE VIADUCTS AND TUNNELS OF THE WHITBY- LOFTUS LINE 61

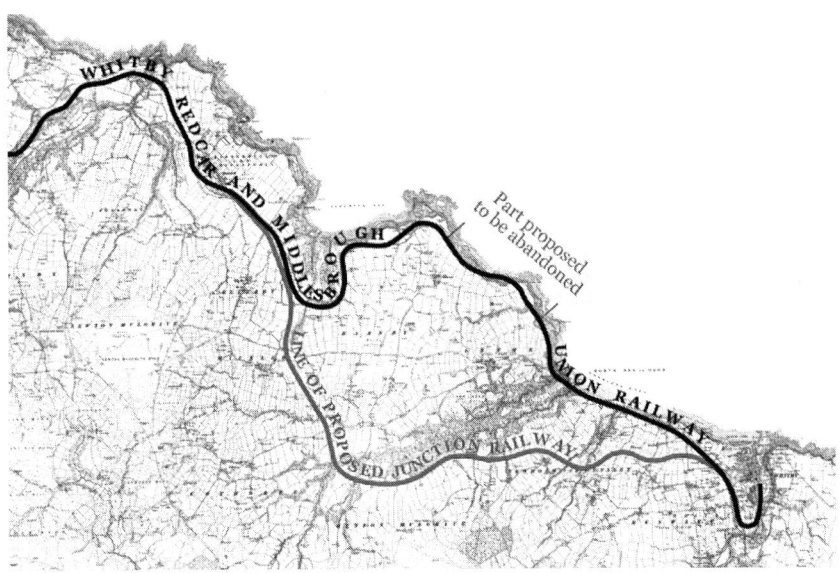

Proposed deviation inland (HL/PO/PB/3/plan 1875/W8).
Forgotten Relics of an Enterprising Age

The design of the viaducts [22]

There was a pleasing, if stark, simplicity in the design which was immediately apparent to all those who saw the viaducts while they existed. *The Engineer* remarked that there was no special novelty in the superstructure, where spans of 60 feet had been adopted. Well-designed lattice girders were thoroughly braced together and these were stiffened laterally by elaborate wind ties which were surmounted by cross iron joists which in turn supported a timber platform 14 feet wide, on which the single line of rails was laid. Because the ravines upon which they were built had at their base shallow streams, there was an equal simplicity in the way in which the foundations of the piers were constructed. The ground was sufficiently solid merely to dig a hole into which the cast-iron bed-plate forming the base of the piers was firmly embedded in a mass of concrete. The piers were constructed of wrought-iron and *The Engineer* seemed very approving of Dixon's idea to fill the piers with cement and concrete, which thus would keep the external skin in shape and dispense with all internal stiffening. 'For the sake of concession to popular prejudice more than anything else', continued the contributor of the article, each pier was equal to a safe load of 300 tons, more than double that which is theoretically required.

Thus the load was not borne by the iron at all 'the real duty of which is to hold the concrete column in shape'. The weight of the column therefore was considerable so the diameter was increased towards the bottom by successive offsets from 2'6" to 3'6" and 4'6" which, for ease of construction, was much preferred to a gradually tapering tube. Finally, weeping holes were punched in the plates at every foot to allow drainage of any excess moisture in the setting of the concrete and each pair of columns was stiffened by diagonal and cross bracing attached to the iron skin of the pier at points duly stiffened to receive it.

It was not the design, but the casual and slack nature of the erection of the viaducts which was to cause so much trouble before the final opening of the line on 3rd December, 1883.

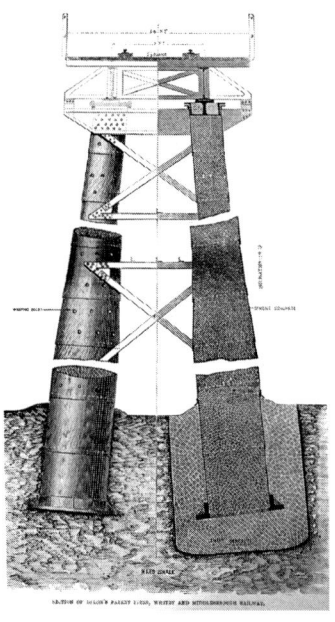

Section plan of Upgang viaduct, *The Engineer*, 14th March, 1873.

Upgang viaduct *c.* 1873, *The Engineer*, 14th March, 1873.

Delivery of the various components of the viaducts

The viaducts were in place quite early. An interesting question arises: how were they delivered? Manufactured at Darlington, the various parts of the viaducts could have been delivered either by rail or by sea. Similar viaducts, with larger spans, were commonly manufactured at that time in works like Skerne's, assembled with temporary bolts at the works, dismantled, sent in smaller pieces to sites and re-assembled and riveted at the site. If this is the case, then the parts could have been transported by rail to Whitby and then moved by road to the four viaduct sites between Whitby and Sandsend. It is, of course, possible that component parts of the viaduct were small enough for them to have been sent by sea and offloaded by the following method. The well-known photographer Frank Meadow Sutcliffe in c. 1887 took a picture of a boat unloading heavy material (coal) on to Sandsend beach, from where it was moved by horse and cart.

Although the process seems arduous, it *may* have been the method of transportation for the component parts of the four viaducts, for they all lay very close indeed to the beach, where it was possible to

Unloading coal on the beach at Sandsend c. 1880.
By permission of Whitby Literary & Philosophical Society, Whitby Museum

ground a boat at low tide. Staithes viaduct, by far the largest of the five, presents a more difficult problem. No material could have been transported from Whitby (Deepgrove tunnel not being completed before 1882) and the nearest railhead was at Loftus five miles away; the Staithes-Loftus section of the line was as problematic as the cliff top section between Sandsend and Kettleness and not completed until July, 1878. How, then, was the heavy material transported to Staithes? It can only have been by sea. There was – and still is – a small harbour at Staithes where it might have been possible for ships, carrying the heavy structures, depending upon their size, from Teesside to have landed. While quite a haul through the steep, narrow streets of the village it was not far to the site. A unique visual representation of the construction of Staithes viaduct may be found in a contemporary text. The illustration shows the viaduct being constructed from the southern side of the valley. The piece being manoeuvred into position by the crane seems small enough to have been able to be carried by ship to Staithes.[23] Lacking any primary sources on the matter one can only conjecture, but, using the principles of Ockham's Razor*, the sea route seems the simplest method of transportation.

Staithes viaduct under construction. *E.M. Hutchinson*

* A problem solving principle which states that a simple solution is more likely to be correct than a complicated one; after the English philosopher, William of Ockham.

Problems

The main source for the history of the viaducts between the takeover of construction of the line in 1875 and its opening in 1883 is the memorandum written by the Chief Engineer of the NER, T.E. Harrison. It took over three years to remedy the defects in almost every area of construction. Then, when things were beginning to improve, the Tay Bridge disaster occurred, leading to, *inter alia*, far greater safety demands on newly built bridges and viaducts. Firstly, in July, 1881 a new set of requirements was issued by the Board of Trade which required that the viaducts should be equal to withstanding a wind pressure of 56 lbs to the foot. The viaducts as built were only calculated to withstand a wind pressure of 28 lbs to the foot. Secondly, a further regulation was made to the following effect: 'If in iron viaducts the main girders are placed below the level of the rails substantial parapets about 4ft 6in. in height must be provided, and as a further protection substantial guards should be fixed outside, above the level and as close to the rails as possible....'[24]

These demands were complied with, but the work was heavy and the cost considerable. Nevertheless by October 1882 the usual notice for inspecting the works with a view to opening the railway was given, and the works were inspected by the Government Inspector, his inspection on that occasion confined to the viaducts, for finding he could not pass them he deferred further inspection on the rest of the line. Far greater detail, however, of this inspection may be gained from the Harrison memorandum (*Appendix One*). It will be recalled that the designer's plan was to fill the piers with cement and concrete, but, wrote Harrison, 'The Inspector required that in the iron casing of the piers of the Staithes viaduct, holes should be cut in them to ascertain the condition of the concrete with which they were filled, the result being that in the first trial the concrete was found to be mere gravel without any cement, and with the same result, though not so bad, in several other cases.' This was clearly a disaster, and perhaps argues more forcefully than any other evidence that the quality of the construction of the line under Dickson and Tolmé was appalling, leading Harrison to remark that 'It has been asked how it occurred that the Engineer-in-Chief had not discovered these defects before. The answer is simply that he did not believe that such scamping of work could take place with even reasonable inspection, and such a case had never come to his knowledge before.' More detailed inspection at this time revealed that several of the piers were not perpendicular, in one case to the extent of 7 inches, in others to the extent of 3 or 4 inches, and this applied more or less to all the viaducts. It was clear that the piers on the tallest viaducts, Staithes and Upgang, had already begun to buckle and remedial work in the form of two rows of longitudinal bracing was necessary. Two rows were, in the end, added to Staithes viaduct and one row to Upgang viaduct. The representation of Upgang viaduct in *The Engineer* article of March 1873 (*page 62*) shows the newly constructed viaduct without such bracing. As well as this, Harrison remarked that upon taking control of the construction in 1875 the roadway for the permanent way on *all the viaducts* was exceptionally flimsy and all had to be replaced. Specifically, this 'roadway' was in the words of *The Engineer* article, 'a timber platform 14 feet wide, on which the single line of rails was laid.'

When this state of the work was discovered it was, understandably, not thought desirable that any formal report should be made by the Government inspector. Next, it was arranged that steps should be taken for a complete examination of every pier in each viaduct with reference to the concrete, and each pier that was out of perpendicular was with great difficulty straightened. No time was lost in remedying these defects and by aid of a force pump machine designed for the purpose liquid cement was forced into every pier.[25]

It is interesting to note that the inspector was Major-General Charles Scrope Hutchinson who in 1878 had inspected – and recommended for opening – the Tay Bridge. According to John Prebble, Hutchinson was 'a stiff, scrupulous, exacting man, who had a sharp eye for the inconsequential detail', and who had only just escaped 'the blame, the anger, and the mob-hatred suffered by Thomas Bouch'[26]

The terrible disaster cannot have been very far from Hutchinson's mind when undertaking later inspections and thus it is likely that he would have made the very highest demands to ensure future

Major-General C. S. Hutchinson.

safety when inspecting the Whitby - Loftus line. Indeed, after the first inspection in July 1883 Hutchinson concluded severely that 'I cannot recommend that the opening of the line should be sanctioned and I must report that by reason of the incompleteness of the works, it cannot be opened without danger to the public using it'. Although not saying so specifically, it was the viaducts which caused Hutchinson the most concern. Hutchinson made 12 requirements to be fulfilled before the line could be opened. Two of these concerned the viaducts: firstly at Staithes viaduct where the longitudinal bracing had to be extended for three spans and the ranging of the girders should be as far as possible improved; and secondly, that in all the viaducts the condition of the concrete required careful examination, and means taken to improve it where defective. Not all was bad, though, for Hutchinson concluded that 'as regards vibration and oscillation, the viaducts now behave well with heavy engines passing over them at speed.'[27] A second inspection followed shortly afterwards, with Major (later Colonel Sir) Francis Arthur Marindin reporting on 22nd August, 1883. But, once again, permission to open the line was deferred; the cause, again, was the defective condition of the viaducts. Indeed, it was clear that the second demand of the previous inspection had not been fulfilled. Major Marindin wrote,

Major (later Colonel Sir) F. A. Marindin.

> With regard to [the second] requirement [of the previous inspection], the concrete filling of most of the piers has been found to be very defective and four of the columns in each of the five viaducts were selected by Maj. Gen. Hutchinson to be dealt with in a manner arranged by him with the Engineer. I examined these columns carefully and I found that, so far as can be judged by sound from tapping the iron casing of the columns, and by inspection at the peep holes and other holes which have been made in the casing, the concrete has now been made good, all the hollow places having been filled up with cement grouting. Orders have been given to continue this work on all the columns, but until all have been satisfactorily healed in the same manner as those which I have tested, I cannot recommend that the opening of this line should be sanctioned and I must report that by reason of the incompleteness of the works, it cannot be opened without danger to the public using it.[28]

However, at last and 2½ years late, the line was approved by the Railway Inspectorate of the Board of Trade. In the final report, made by of Maj. Gen. Hutchinson on 3rd November, 1883, he gave permission that the line be opened, but it was apparent that he considered that the viaducts still had the capacity to cause problems. On each viaduct the maximum speed allowed was 20 mph. Since Major Marindin's report he found that the concrete in the columns of the viaducts has been carefully gone over, and its defects made good, except in three places in Nos. 5, 6, and 8 piers at Staithes, which were to be at once attended to. He made a number of further demands, firstly that the viaducts be carefully maintained particularly as regards the bracing, both longitudinal and transverse. Secondly, that no time should be lost in painting those portions of wrought-iron girders which had not yet been recently painted (this was an ongoing task throughout the line's history). By far the most unusual demand of Maj. Gen. Hutchinson was that, at Staithes, because of its height and exposure to easterly gales, a wind gauge should be placed in a suitable position, in charge of the Staithes station master, and that no train should be allowed to cross the viaduct when the wind registered a force of 28 lbs to the square foot or more.[29]

Staithes viaduct wind gauge, it is now in the National Railway Museum, York.

J. W. Armstrong/ Armstrong Photographic Trust

> **NORTH EASTERN RAILWAY.**
>
> General Manager's Office,
> YORK, March 18th, 1884.
>
> ## REGULATIONS AS TO STAITHES VIADUCT,
> ### WHITBY AND LOFTUS LINE.
>
> Engine drivers and guards are required to note that under no circumstances must engines or trains travel along this viaduct at a higher rate of speed than 20 miles an hour :—
>
> No train must be permitted to travel along the viaduct when the wind-gauge affixed to the viaduct registers a force of 28 lbs. to the square foot or more.
>
> The wind-gauge is so connected electrically with a bell in the signal cabin that should the wind pressure amount to 28 lbs. per square foot the bell will ring continuously so long as that or any higher pressure exists.
>
> When the bell commences to ring, the signalman must at once cancel any "be ready" signal he may have accepted for a train from the north, and if the train has been accepted and is "on line" he must put on both home and distant signals against it, and inform the Station Master, who must at once take all practicable measures to stop the coming train before reaching the viaduct. If a train should be coming from the south it may be permitted to come up to the station.
>
> A telegram must, if practicable, be sent to the District Superintendent informing him of the circumstance whenever during a storm the bell commences to ring, so that he may make any special arrangements he may deem necessary ; and full written reports of each such case, giving the time during which the bell rang, and the trains were detained, must be transmitted to the District Superintendent as soon afterwards as possible.
>
> HENRY TENNANT,
> *General Manager.*

(TNA RAIL 527/908). Regulations as to the Staithes viaduct and wind gauge.

Extremely detailed instructions were issued by the NER three months after the opening of the line concerning the operation of the railway in windy conditions:

Anecdotal evidence suggests that these instructions may, in later days, have been honoured more in the breach than the observance, owing to the inadequacies of the gauge.[30] No doubt there were sighs of relief in both the NER and WRMUR boardrooms but, as Harrison reminded them just after the opening of the line, 'the amount spent in putting the viaducts alone into a proper state exceeded £30,000, but for the work to the viaducts the line would have been completed nearly 18 months earlier.'[31]

The dimensions of the viaducts

In the Board of Trade inspector's report of July 1883 Maj. Gen. Hutchinson went into considerable detail when discussing the dimensions of the five viaducts. There was considerable variation in the height and length of the viaducts: [32]

	Height	Maximum span	Girder type	Length
UPGANG	70'	60'	Lattice	330'
NEWHOLM	50'	30'	Plate	330'
EASTROW	30'	60'	Lattice (6 spans); Plate (2 spans)	528'
SANDSEND	63'	36'	Plate	268'
STAITHES	152'	60'	Warren (6 spans); Plate (11 spans)	790'

The line opened, beginning its 75 year life quietly.[33] As for the viaducts, they initially gave no problems but, by 1895 (only 12 years after its opening) the lattice girders at Upgang viaduct required strengthening and at Eastrow viaduct complete replacement (owing to corrosion, the viaduct being built actually on the beach and within 35 feet of the sea at high tide). Certain struts on Upgang viaduct had buckled considerably and needed urgent attention while at Eastrow it was decided to replace the lattice girders with new steel girders. The contractors were the Cleveland Bridge and Engineering Company of Darlington. The contract for the Upgang viaduct was entered into on 6th December, 1893 and the work finished by 30th August, 1894; the contract for the Eastrow viaduct was entered into on 9th May, 1895 and the work was finished by 31st October, 1895. The cost of the work reflected the seriousness of the nature of the problems: repairs at Upgang cost £714, while those at Eastrow cost £2,682 14s. 9d.[34] The viaducts were regularly painted and maintained and gave no further problems.

History is silent concerning the viaducts until the announcement by British Railways in September 1957 that the line would be closed, stating that the main reason for closure was the uneconomic nature of the line and that £57,000 would be required for maintenance of the tunnels and viaducts over the next five years. Despite a rearguard action being fought by the Whitby Urban District Council under the leadership of the Clerk, Mr J.B. McClurg, the line closed to all traffic on 5th May, 1958, the last train running on 3rd May, 1958.[35]

Demolition and destruction

Two years after the closure of the line in 1958 demolition of these remarkable structures began. There is considerable photographic evidence for this process. Perhaps it is instructive to contemplate the ease and brutality of their destruction with the difficulties involved with their construction. Indeed, the second Prospectus issued by the WRMUR described the Staithes viaduct as 'noble'.

Unfortunately, it has not been possible to locate any photographs of Upgang viaduct in process of demolition. The viaducts were cut up for scrap.

Two photographs of the demolition of Staithes viaduct. *British Rail*

Two photographs of the demolition of Sandsend viaduct. *K. Hoole*

THE VIADUCTS AND TUNNELS OF THE WHITBY- LOFTUS LINE 73

Two photographs of the demolition of Eastrow viaduct. *K. Hoole*

The demolition of Newholm viaduct. *K. Hoole*

The tunnels

As finally built there were three tunnels on the 16 mile line between Whitby and Loftus. These were the Deepgrove (or Sandsend), the Kettleness, and the Grinkle (originally Easington) tunnels.

The longest was Deepgrove tunnel (1,649 yards [1,508 m]). The tunnel is, for the most part, straight; however, there is a curve to the north for the last 300 yards.

However, the only tunnel authorized in the original 1866 Act was the Easington, which was then to be 1,324 yards long, when work on the line began with the cutting of the first sod on 25th May, 1871. Progress was satisfactory along some sections of the line, but the Engineer's Report presented to the Directors of the company at their meeting of 9th May, 1872 stated baldly that on the Easington tunnel to Loftus section, 'Nothing has been done. Machinery there, then found the tunnel could be halved by a slight deviation. Careful examination of the ground satisfies us that we shall not meet with any extraordinary difficulty in the execution of the tunnel.'[36] The Engineer, Mr J. H. Tolmé, was often given to exaggeration and prone to telling

THE VIADUCTS AND TUNNELS OF THE WHITBY-LOFTUS LINE 75

Deepgrove (Sandsend) tunnel (south portal). *K. Hoole*

Deepgrove (Sandsend) tunnel (north portal). Note the distorted brickwork on the left-hand side of the portal. *N. Cholmondeley collection*

The north portal of the 308 yards (282 metres)-long Kettleness tunnel.   *C.C. Cobb*

Kettleness tunnel (south portal). Note the check rail on the tight curve.
N. Cholmondeley collection

the Directors what he thought they wanted to hear. His forecast concerning the future construction of the Grinkle (Easington) tunnel was incorrect. The tunnel was shortened, though, for Tolmé's report of 9th September, 1872 mentions that, after arrangements with the landowners, the tunnel has been shortened from nearly a mile in length to considerably under half a mile (792 yards).[37]

This deviation, along with other changes to the 1866 plans led the Directors, at a meeting on 12th October, 1872 to note that 'it is desirable that parliamentary sanction should be obtained to the deviations which have been necessary in carrying out the works upon the railway'.[38] At the Directors' meeting of 11th February, 1873 they 'resolved that the Bill now submitted for authorizing the diversion and alteration of the line and levels of the WRMUR and for other purposes be and the same is hereby approved'.[39] It is interesting to note that work had already been undertaken on these alterations, the main ones being bringing the line closer to the sea between Whitby (West Cliff) and Sandsend and the tunnelling changes at Grinkle and along the cliff face between Sandsend and Kettleness. Thus the 1873 Deviation Act was passed.[40] The plans issued with the 1873 Act clearly show the new length of tunnel at Grinkle and six new ones proposed between Sandsend and Kettleness.[41] However, there is no mention of any moneys spent on tunnelling until the 21st Engineer's certificate presented to the 4th April, 1873 meeting of the Directors where it was noted that £1,000 had been spent on tunnelling.[42] This must have been on the Sandsend-Kettleness section where, Tolmé reported to the Directors on 1st September, 1873 'the headings for the short tunnels through the jutting points of the cliff are nearly all driven.' Unfortunately, in the same report, Tolmé had some bad news concerning the Grinkle tunnel: 'A slip which took place in the embankment at the southern end of the Easington tunnel has caused a great deal of trouble and delay....A bed of quicksand was also met with in the tunnel heading, which considerably retarded the work....'[43]

Without any doubt at all, the decision to initially construct the line *around* the cliffs between Sandsend and Kettleness was not only a major mistake, but the main cause of the line's failure. Nevertheless a considerable amount of work was done on the various tunnels as is indicated by Mr Tolmé's last engineer's certificate. In that certificate of 20th October, 1873 he reports that, so far, £4,681 had been spent on the various tunnels. The problems caused by the cliff edge works as well as those at Grinkle tunnel brought the company to the edge of bankruptcy, and resulted in the sacking of the contractor, John Dickson at the end of

The south portal of Grinkle tunnel. It was originally called Easington tunnel and was 990 yards (905 metres) long. *D. Ibbotson*

Grinkle tunnel (north portal). *D. Ibbotson*

1873 and of the Engineer a few months later. It is impossible now to know how much work was actually done (no earthworks of any kind remain on the cliff edge section). However, a more recent historical Ordnance Survey map (before 1923) shows a tunnel at Keldhowe Point on the abandoned section of the line. It is astonishing to think that a railway line could ever be constructed around these cliffs – but it was, although the rails were never laid. Not only that, the plans also demanded that between Keldhowe and Kettleness there should be six tunnels built (*see the plan page 120*). The longest of these, 154 yards long, was to be built below the headland of Keldhowe.

Today there is very little visible evidence to show that any major construction occurred, for the cliffs are unstable and there are regular landslips and rockfalls. It seems unlikely, if not impossible, that a tunnel could be built here, but there is incontrovertible evidence to show that it was. As well as the O. S. map (the old and new lines have been superimposed on the map, *see page 80*), there is also a letter, written from the Royal Hotel in Whitby and dated 2nd December, 1873. It is written by the son of the Secretary of the company which was building the line, to his father. The son, George Fraser, had been sent to Whitby to oversee the building of the line which was going through considerable difficulties at the time. Part of his work was to travel over the entire line to see how the work was progressing. Travelling to the cliff edge he wrote, 'I was struck yesterday in going over the works of their very heavy nature. The cuttings along the face of the cliffs (i.e. at Keldhowe) must have cost a vast amount of labour. But it seems to me that there were not half enough men employed there. Fancy, for instance, *ten* men working their utmost at the masonry of the cliff tunnel and perhaps ten more excavating the tunnel itself. I don't know when the tunnels would be finished at this rate.' Five months later (in May 1874) another visitor to the area wrote, '….the railway although grand if successfully completed along the cliffs appears to me to bid defiance to safety; already thousands of tons of rock are blocking the line….'

So dangerous was this construction that it is difficult to believe that the cliff-top plan could be taken seriously. Yet it was, and the building of the cliff-edge line continued until the original company (The Whitby, Redcar and Middlesbrough Union Railway Company) went bankrupt in 1874 and was forced to come to an agreement with the North Eastern Railway Company in which the latter company agreed to complete the line. The Whitby Company was always under-funded and struggled to attract investors. The first thing the North Eastern company did was to abandon the cliff-edge route and plan the inland tunnels which exist today.

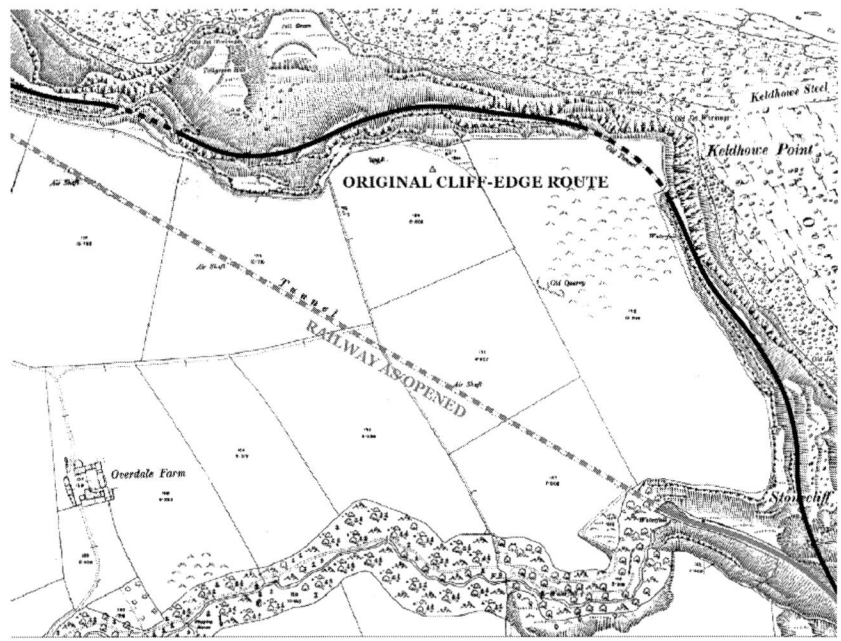

Ordnance Survey map (pre-1923) showing 'Old Tunnel' at Keldhowe Point. The black line indicates the abandoned section; the grey line as finally constructed.

Forgotten Relics of an Enterprising Age

However, even the major tunnelling works which eventually took place further inland did not lessen the cost.

By the summer of 1874 the company was in considerable financial difficulty. Creditors were threatening to take legal proceedings against the company; horses and locomotives belonging to the company were sold at auction; complaints from members of the public regarding damage caused by the company and letters from disgruntled shareholders were arriving regularly; attempts to attract further investment failed dismally[44] and, by 12th November, 1874, the company's bank account stood at £963 15s. 6d. [£44,054 today] (and this only after the better of the two locomotives had sold for £750).[45] The contractor had been sacked, and work on the line had ground to a complete halt. There was only one option open to the Directors if the line was to be completed: an agreement with the North Eastern Railway company. On 12th November, 1874 the Directors met to a consider a letter they had received from Mr C. N. Wilkinson, the Secretary of the NER, dated 23rd October, 1874 in reply to the WRMUR's Directors' wish to open negotiations with the NER for taking over the construction of the line. At this moment the NER's reply was cautious:

> Dear Sir Harwood,
>
> The representations which have been made as to the present position of the WRMURly have occupied the attention of the North Eastern Board and the decision that they have arrived at is that they *cannot* [my italics] advise their shareholders to adopt the *present* line with a view to its completion or to become subscribers to it. The N.E. Board however are prepared to look *favourably* on the construction of a line of railway properly laid out to meet the requirements of the district and further, so far as can be done, to adopt such portion of your present line as will be consistent with the construction of such new line and to utilise so far as possible any materials, works, etc. which may be available for the new line *paying for the same* according to a fair valuation. If your company thinks it advisable to enter into negociations [sic] on this basis you will of course advise the Directors of this company.
>
> Yours faithfully,
> C. N. Wilkinson (Sec.)[46]

Central to the demands of the NER was that the cliff edge line be abandoned; clearly it was not, in their opinion 'properly laid out'. At last in May 1875 an agreement was made, ratified by the Whitby, Redcar and Middlesbrough Union Railway Act of 19th July, 1875, between the North Eastern Company and the Directors of the Whitby, Redcar and Middlesbrough Union Railway to complete the line 'with all despatch', and 'in a substantial and satisfactory manner'. The NER would find the necessary capital and work the line upon completion.

Initially, the idea was for a major deviation away from the coast to be made. This would certainly have been a cheaper and easier option, but the new line was never begun. Interestingly, it would have included a tunnel near Mickleby of 525 yards in length at an estimated cost of £13,650. The cliff edge section between Stonecliff End and Holmsgrove was indeed abandoned, but the deviation line inland was never built (*see map page 61*); instead, a tunnel of length 1,652 yards was built which by-passed the cliff edge line and ran from Deepgrove to Holmsgrove where it emerged for a short section in the open air before entering another – Kettleness – tunnel. It is likely that the North Eastern Railway (with, no doubt, the agreement of the WRMUR) considered that too much had already been invested in the line to make such a major deviation, or, more likely, that the additional new cost would not be recouped by savings in working or maintenance.

The Directors' Report to their shareholders of 22nd February, 1876 clarified the position:

> As provided by the terms of the Agreement, the North-Eastern Company has applied in the present session for powers to make a Deviation of the authorised line, and the Bill is about to be submitted for your approval. The deviation is less extensive than was proposed last Session, and substitutes for the line along the cliff about a mile and a quarter of railway almost all [sic] in tunnel.[47]

On 13th July, 1876 powers for the deviation of the line through what is now known as Deepgrove tunnel (and for the abandonment of the partly constructed works along the cliff top) received the Royal Assent.[48] The North Eastern Railway had immediately recognized the main problem, but its expensive solution (the construction of Deepgrove tunnel) was to become an important contributory factor for the line's financial difficulties throughout its life and, indeed was, the immediate cause of its closure.

However, between 1875 and 1879 no work on this new tunnel was undertaken. The reason for this is clearly explained by the memorandum of the NER Chief Engineer, Mr T. E. Harrison. Mr Harrison was writing just after the long-delayed opening of the line in December 1883 and felt the need to explain and justify the length of time it had taken to construct such a relatively short line. According to this memorandum almost every element in the construction of the line when taken over by the NER in 1875 was faulty. This document has been discussed elsewhere and it is a most important source for understanding the early history of the line.[49] Concerning the tunnels Harrison wrote,

> ....the surveys were so inaccurate that had the original tunnels been completed [those around the cliffs and at Easington], they were so out of line with each other, that no junction could have been made with them at all.[50]

The Harrison memorandum gives the clearest explanation as to why the construction of the Deepgrove tunnel began so late in the proceedings. It was necessary to get all the other deficiencies remedied first. At Grinkle tunnel problems continued; the Harrison memorandum indicating that the delay in opening the line was caused in part by two major landslips at the south end of the tunnel and another a little further south 'extending over four acres of land which slid away for 300 yards.'[51]

Reports submitted to the half-yearly meetings of shareholders between October 1879 and October 1882 indicate that this was the period of the construction of the tunnels between Sandsend and Kettleness. On

THE VIADUCTS AND TUNNELS OF THE WHITBY- LOFTUS LINE   83

Photographs of construction in progress at the south end of Grinkle (Easington) tunnel.
By permission of Whitby Literary & Philosophical Society, Whitby Museum.

Friday, 24th October, 1879 the shareholders of the company were informed that 'The Mulgrave [i.e. Deepgrove] tunnel is commenced, also involving in its construction a large amount of very tedious work; but the Directors have reason to believe that the Contractor is proceeding surely and steadily (about 800 men being daily employed....)'[52] Six months later the shareholders were informed that, 'Much time, however, must yet elapse before the line can be opened throughout, owing to the very heavy amount of tunnelling to be accomplished....'[53] Another optimistic report in October 1880 indicated that, 'The Directors are informed that rapid progress is now being made....where the tunnel through the ironstone beds, near the Alum Works, is being vigorously pushed forward....[54] However, it was at the next shareholders' meeting that the problem of landslips on the line become serious. At this meeting (17th March, 1881) a letter from Harrison was presented in which he gave details,

> The material at the Easington tunnel had for some distance turned out to be wet sand instead of shale as was expected, and this requires the tunnel to be invested with stronger brickwork and all to be timbered in the tunnelling, which as near as I can estimate will cost about £10,000 more than the contract.[55]

Construction in progress at the south end of Kettleness tunnel.
By permission of Whitby Literary & Philosophical Society, Whitby Museum

The line around the cliffs at Kettleness proved equally problematic. Harrison recalled in his 1883 memorandum that,

> The three great slips that have taken place on this line are such I have never before in my long practice experienced....The slip in the face north of the Kettleness Tunnel has rendered it necessary to make a considerable deviation of the line for a length of 726 yards, including a tunnel 308 yards in length....The Kettleness [sic, he means Deepgrove] Tunnel is 1,621 yards long; a heading is made throughout its whole length, and there are only 540 yards remaining to be completed, 1081 yards being finished....[56]

This work proved costly: Kettleness tunnel cost in the region of £5,000 to construct. The tunnelling work was still in progress as late as February 1882. A complaint by the master of a ship that had run aground near Whitby because lights on the cliffs near Sandsend had been mistaken for the Whitby lighthouse was investigated by Trinity House. Coastguards at Staithes and Whitby stated that these 'lights' were in fact open fires at the entrances to the long tunnel at Deepgrove and Holmsgrove, provided for illumination for the shift workers involved in the tunnelling. This had been going on for two or three years.[57] After the complaint was dealt with these lights were discontinued. It seems that the tunnels had been completed (or nearly completed) by the October of 1882, for the Shareholders' report for that month stated that, '....general engineering works....nearly completed....extra works...required on the viaducts....'[58]

The North Eastern Railway spent a considerable amount of money on the construction of the three tunnels on the line. The final cost for the line amounted to £655,077 of which the NER contributed well over £350,000. This high cost was due almost entirely to the need to improve the viaducts, to build or complete the tunnels and to make all the improvements so clearly delineated in the Harrison memorandum. Nevertheless the final product reflected the money spent upon it. Visual evidence, showing the interior of the Sandsend (Deepgrove) tunnel over 120 years after its construction and 50 years after its abandonment shows a very solid and well-built structure. Indeed, so well-built were the tunnels that there is very little reference to them in the three formal inspections carried out by the Board of Trade in 1883, thus indicating that they were not a cause of any concern for the inspectors.[59]

The development, construction, and final completion of the line were by no means easy. The main problems concerned the topography through which the line passed and the difficulties encountered in

Inside Deepgrove tunnel, 2011. *Author's collection*

overcoming those problems. The – in retrospect – foolish decision to round the cliffs between Sandsend and Kettleness and the immense costs involved in trying, first of all, to implement that decision and then the necessity to construct two tunnels were factors which were to influence the consequent history of the line, even up to its closure. Problems with the Easington tunnel were also costly.[60]

There were two slightly unusual elements in the construction of Deepgrove tunnel. Two adits or spoil tunnels were constructed so that some of the excavated material could be taken right to the cliff edge and easily dumped onto the rocks below.

The line was difficult to work (*see gradient map page 30*). Especially difficult for train crews was the 1 in 57 bank from Sandsend to Kettleness (with a short stretch of 1 in 62). Most of this steep climb was through Deepgrove tunnel which had been built with two ventilation shafts. The line had been operating for 16 years before it was decided to ease the difficulties for the drivers and firemen labouring up this difficult stretch of line. Although there is no direct evidence, complaints from the train crews must have been regular enough and loud enough to make the NER construct two additional shafts. Detailed plans of these exist in the National Archives.[61] They were built in 1900.

THE VIADUCTS AND TUNNELS OF THE WHITBY-LOFTUS LINE 87

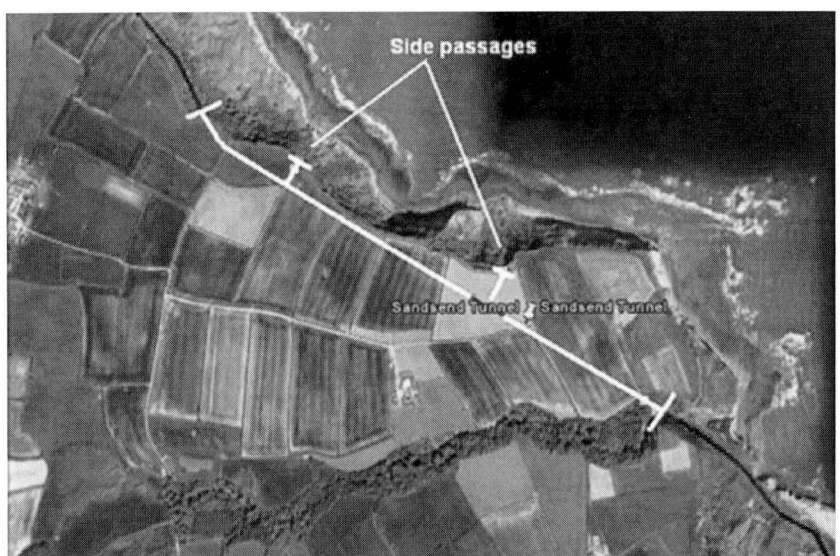

Aerial photograph showing superimposed tunnel and adits. *Author's collection*

Deepgrove tunnel adit. Excavated material was taken to the exit and dumped onto the rocks below. *P. Davison*

Looking up one of the ventilation shafts (built 1900) in Deepgrove tunnel.

Author's collection

As far as the primary sources are concerned, the tunnels pass out of history until September 1957, when the announcement was made that British Railways were to close the line.[62] The ostensible reason was the cost of immediate and future maintenance, it being considered that £57,000 needed to be spent, mainly on the tunnels and viaducts, in the next five years. This, plus the limited income from the very few passengers that the line carried, was a difficult argument to counter. A glance at the north portal of Deepgrove tunnel, (*see page 75*), shows the distortion of the brickwork and the out of alignment tunnel mouth. This photograph was taken *c.* 1956. British Railways were sufficiently alarmed by the inherent danger of the situation that they put in place temporary repairs, reinforcing the misshapen brickwork with steel rails.

However, while maintenance problems were absolutely necessary, it is more likely that the demands of the 1955 Modernisation Plan had more to do with the closure of the line than the (admittedly large) expense of maintaining the tunnels and viaducts.[63]

When the line closed to all traffic on 5th May, 1958 (the last trains running on 3rd May) the tunnels were left to the vicissitudes of time and

THE VIADUCTS AND TUNNELS OF THE WHITBY-LOFTUS LINE

Taken 50 years after the closure of the line, this photograph clearly shows the attempts made by British Railways to temporarily shore up the north portal of Deepgrove tunnel. It was the extent of these problems which led to the announcement that the line was to be closed. *Author's collection*

weather. However, they were so well-built that well over 50 years later the interiors of Kettleness and Deepgrove tunnels are still in good condition, except for the collapse at the north portal of Deepgrove.[64] Access to the south portal of Deepgrove tunnel is straightforward, as is the north portal of Kettleness tunnel, but the respective north and south portals of these tunnels can only be reached by passing through the tunnels, as the trackbed on the cliff face at Holmsgrove is totally inaccessible. While the ventilator shafts of Deepgrove tunnel exist, the exits in the fields above have been covered over, and there is now no sign of them.

The north portal of Kettleness tunnel today.
Author

The south portal of Kettleness tunnel today.
Author

The south portal of Deepgrove tunnel today. *Author*

The north portal of Deepgrove tunnel on the verge of collapse. In 2008 there was a major collapse. *Author's collection*

CHAPTER FIVE

A BRIEF FINANCIAL HISTORY OF THE LINE FROM 1897 TO 1940

Introduction

The agreement between the NER and the Whitby Company, while allowing the line to be completed, put the Whitby Company in a very weak position and forced it to sell out to the former in 1889 at a very considerable loss. Nevertheless the fortunes of the line slowly began to improve as shown by the very detailed traffic receipts surviving for the years 1897-1907[1], 1910-37,[2] and 1938.[3] Evidence for the years 1939 and 1940 is less detailed (*see page 113*). The 1897-1907 receipts concern all the seven stations on the line and contain the following information: the year, the number of passengers booked, the passenger receipts, receipts from the transport of parcels, horses, carriages, dogs and other (un-named) miscellaneous items, the total coaching receipts (i.e. those from both passengers and parcels, horses, carriages and dogs), the numbers of livestock (both in and out), tons of coal, coke, lime and limestone (in and out), tons of goods forwarded, tons of goods received, the expenses of the station, and the total receipts for the year.

The traffic receipts for the years 1910-37, also for all stations, are even more detailed. Compared with the 10 elements detailed in 1897-1907, there are 15. These are: the number of passengers booked, the number of tickets collected local to the station, the number of tickets collected beyond the station, the number of parcels forwarded, the number of parcels received, the total number of parcels, passenger receipts, receipts for parcels, horses, carriages, dogs and other un-named miscellaneous items, total coaching receipts, livestock (number of heads in and out), coal, coke, lime, limestone and ironstone received (in tons), goods forwarded (in tons), goods received (in tons), the station's expenses (salaries and wages) and the total receipts at the station.

The 1935-7 receipts are equally as comprehensive as those for 1910-34, but show slight differences. The 14 elements are: the number of passengers originating (total tickets issued and total journeys), the number of journeys made annually on season tickets, the number of tickets collected (both to and beyond the station), the number of parcels received and forwarded and the total number, other merchandise received and forwarded in tons, gross receipts for ordinary passengers, season ticket holders and parcel and miscellaneous traffic, miscellaneous other receipts, and station debit. The 1938 returns are equally detailed, the headings being: parcel and other merchandise traffics charged at passenger train rates, coaching traffic receipts, miscellaneous receipts (e.g. lavatories, pillows and rugs, cloak room), tickets collected, passenger traffic (e.g. standard or ordinary Fares, monthly returns, workmen) and season and traders' tickets.

These traffic returns tell a remarkable story and provide evidence for the argument that the line was not always unsuccessful; indeed, for a very few years it was quite the opposite, and then suddenly, dramatically, and catastrophically after the First World War passenger numbers fell to such a low level as to, in most cases, halve the total station receipts (this is particularly noticeable at Sandsend). The evidence also shows that certain elements of income did maintain a reasonable level (although certain anomalies, especially at Loftus, are apparent).

The earlier period (1897-1907) shows a slow but steady growth in income, and this, allied with an even stronger growth up to the mid-1920s, allows some questions already asked in earlier chapters, to be answered. The traffic returns, along with United bus timetables for the area, provide the main sources of evidence.[4] The statistics show more powerfully than any other source how the introduction of the motor bus to the area affected the stations along the line, except, perhaps, for Kettleness, the only station which could be considered remote, and far from any major road which carried public transport. The other compelling piece of evidence is that which shows how successful the two years 1919 and 1920 were. The numbers of passengers booked at each station leap dramatically, and only fall away to more or less the levels seen before the sudden increase for three or four years afterwards until the shocking decline which begins in 1925 and continues remorselessly until the line's final closure.

How much did the line contribute to rural mobility? It will be argued later in this chapter that the contribution was considerable. Can Irving's conclusion that 'the line was a financial disaster of some magnitude' be extended to the period 1890-1934? The evidence of the traffic returns is enough to show that this assessment would be too harsh. Indeed, some of Irving's more general conclusions concerning the NER may be called in question. He maintained that 'after 1901 the passenger traffic ceased to display the vitality characteristic of the 1880s and 1890s'. This is patently not the case on the Loftus-Whitby line and, indeed, for a few years (if only a very few) Irving's assessment that the line was 'marginal and unprofitable' is incorrect. Indeed, Irving admits that 'the branches (including the Whitby-Loftus line) were not moving handfuls of people. They moved large numbers, thus reminding us of the social function provided by railways for isolated communities ... while they (branch lines) may not have made a great deal of money, at least some of the lines were quite busy and provided a genuine service for their local communities'.[5] Should the line have been built at all? Again, the traffic returns for the period show that the line provided a very valuable service for both freight and passenger traffic in this period, thus justifying its existence, at least until the early 1920s. Thereafter the case is not as strong, particularly in terms of passenger traffic after *c.* 1925.

A question that has not been considered so far is the line in the context of bus services. Although there are no traffic returns for the United buses, the sudden fall in certain classes of passenger-related traffic indicate that the buses were probably the prime cause of the disaster (and it is not too dramatic to use such language) which befell the line after *c.* 1925. The United bus timetables, especially those in the middle and late 1920s show the quick development of the network in the area. There is also evidence to show how the London & North Eastern Railway (LNER) reacted to the growing road competition, by creating cheap-day return tickets which were often lower than bus fares. That the line continued to have an important social value is also apparent from the traffic returns, especially before the incursions of the United Bus Company. As to who used the line: this is more difficult to assess, although an attempt will be made to answer this question in the course of the chapter. So how did the line manage to survive the catastrophic fall in passenger numbers? Was the line's existence for a further 30 years based mainly on its social function rather than any other element, or simply managerial conservatism?

To what extent did the fortunes of this line provide evidence for the argument that the financial problems of the railways in the 20th century had their origins in the late 19th century? Leunig's 2006 article, one of the most recent in a strand of econometric analysis now over four decades old, argued that although British Railways benefited the economy and society in general, they benefited the passengers particularly. It reinforces the argument that the 19th century railways' social function was as important as their profitability.[6] It was the *time* saved by railway travel that was crucial: trains were much faster than coaches or walking; the cost of travel fell significantly, and its speed and comfort rose dramatically. As regards society in general, Leunig wrote that, 'The form of cost-benefit analysis used by historians to study railways is known as 'social savings'. Put simply, the social saving from railways is the minimum additional amount that society would have to pay to do what the railways did without them, that is, the cost of moving freight and passengers without trains'.[7] Economically, Leunig thought that 'the number of hours saved (that is, hours that could be spent at work) rose dramatically over the railway era, from 54 million hours in 1843, to 527 million hours in 1866, and finally to 5 billion hours by 1912, roughly equal to the hours worked by 1.8. million workers'.[8] Thus the railways (and here we must remember to include the Whitby-Loftus line) were part of a major transformation of British economy and society, so much so that, according to Leunig, 'The social savings from railways in time and money amounted to some 2 per cent of GDP as early as 1850, to 5 per cent of GDP by 1865, 10 per cent by the turn of the century, and fully 14 per cent by 1912'.[9] As for benefiting the passengers (both the middle and working classes, for third-class accommodation was used by many middle class passengers, certainly by the late 19th century), Leunig made the telling comment that 'people who could never have expected to travel in all their lives were able to do so for the first time. And those who did travel were able to do so more often'.[10] This was especially the case from the 1870s onwards, as railways sought to attract more customers with better third-class services (and, although Leunig does not say so, by implication the building of new lines). The railways may have offered poor returns to investors, concludes Leunig, but they delivered tremendous welfare gains to travellers and to society.[11] The Loftus-Whitby line, then, can just as easily be appreciated in a positive manner, in Leunig's terms, as in Irving's much more pessimistic view.

A final observation may be made from a study of both the available train and bus timetables from the period: there seems to be no attempt whatsoever at any rail/bus integration, even once the LNER had secured a financial interest in local bus operators.

**The position of the stations;
their proximity to the towns and villages;
evidence from traffic statistics**

The position of the station *vis-à-vis* its town or village was an important one, especially after the introduction of regular bus services and the concomitant threat to the fortunes and relevance of the railway. Whitby (Town) station, although strictly not a part of the WRMUR, was nevertheless the starting point for services before the opening of the Whitby (West Cliff) to Scarborough line in 1885 and, indeed, was so used for some services up until closure in May 1958. The Town station was – and is – ideally placed. It lies in the centre of the town, close to the tourist attractions, the main shopping area, and, until the expansion of

A BRIEF FINANCIAL HISTORY OF THE LINE FROM 1897 TO 1940   97

the town which arguably began in the 1920s, to most housing. The bus station is adjacent to the railway station (it opened on 6th April, 1939) and, if there is any evidence at all of bus/train integration, this is the only example that can be cited. During the period of the Whitby-Loftus line's existence there were considerable sidings to the east of the station.

Whitby (West Cliff)

Whitby (West Cliff)'s position was favourable, but only for tourism. Maps of the area indicate that the locality was not built up to any degree before *c.* 1925. It was a short walk, less than five minutes, from the station to the built-up holiday area of the west cliff and thence down to the old town. In its later years the station served a new and well-developed area, as indicated in the post-war O.S. map (*see page 10 lower map*), but by then the omnibus had become dominant.

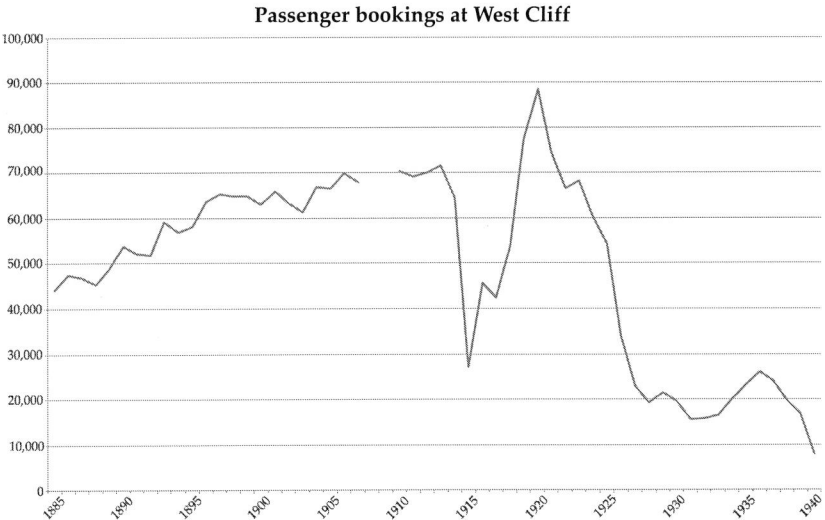

Sandsend

Of all the stations to suffer from the incursions of the omnibus, Sandsend showed the greatest decline. This will be discussed in detail later in the chapter, but the development of a new and fast road between

Whitby and Sandsend in 1925, and the parallel development of local bus services explain the catastrophic fall in passenger usage. Sandsend, almost entirely part of the estate of the Marquis of Normanby, stretches along the coast road for over half a mile to include the district of Eastrow. The station was built above and to the west of the village, certainly close to the housing, but involving a steep climb of 50 yards or so up Lythe Bank to the station entrance. For the purposes of tourism, the station was in an ideal position, being very close indeed to the beach.

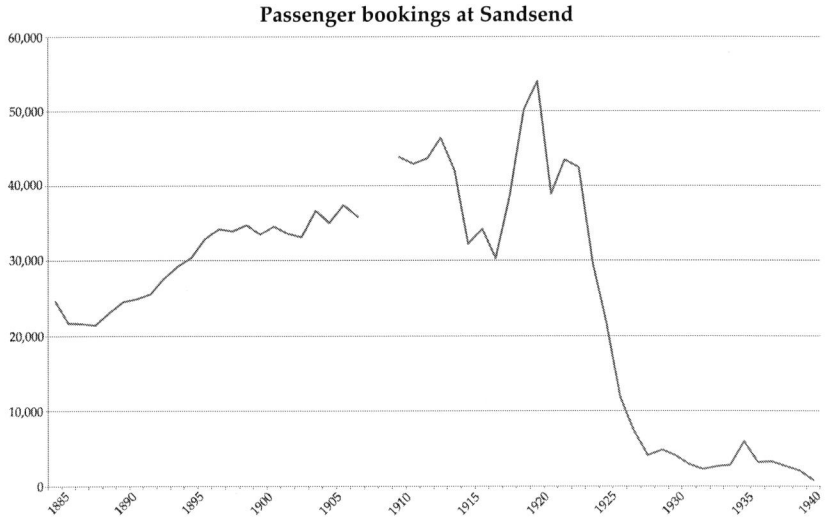

Kettleness

Kettleness station was the remotest on the line. It was some distance from a main road (all the other stations except Grinkle were not) and there were no nearby villages (*see map page 19*). Consequently the receipts were relatively low although, they did not fall away as drastically as was the case with the other stations on the line.

Hinderwell

Hinderwell station, however, served a large village – almost a small town – and was not inconveniently placed (*see map page 21*). It also provided the station for the nearby picturesque coastal village of Runswick Bay which was, and still is, popular with tourists. A station for

A BRIEF FINANCIAL HISTORY OF THE LINE FROM 1897 TO 1940   99

Runswick, 700 yards or so to the east of Hinderwell station was proposed, and detailed plans were drawn up for its possible construction, but nothing came of this, possibly because it was too close to Hinderwell station[12] (*see Chapter 10*).

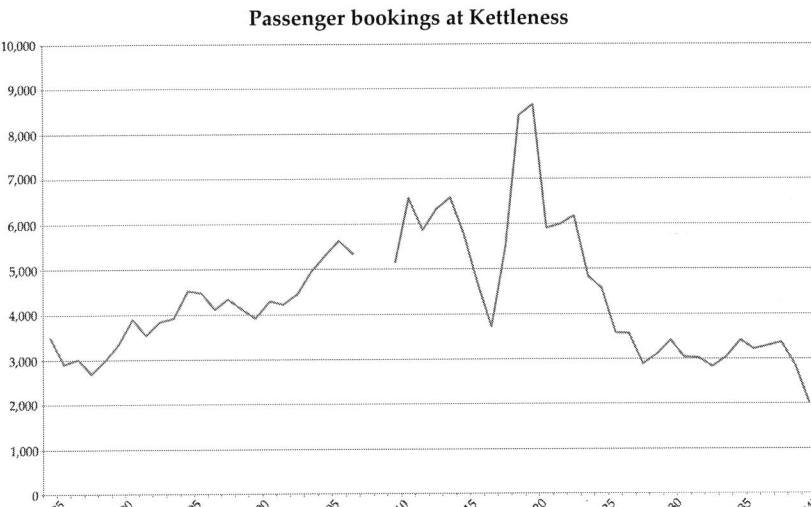

Passenger bookings at Kettleness

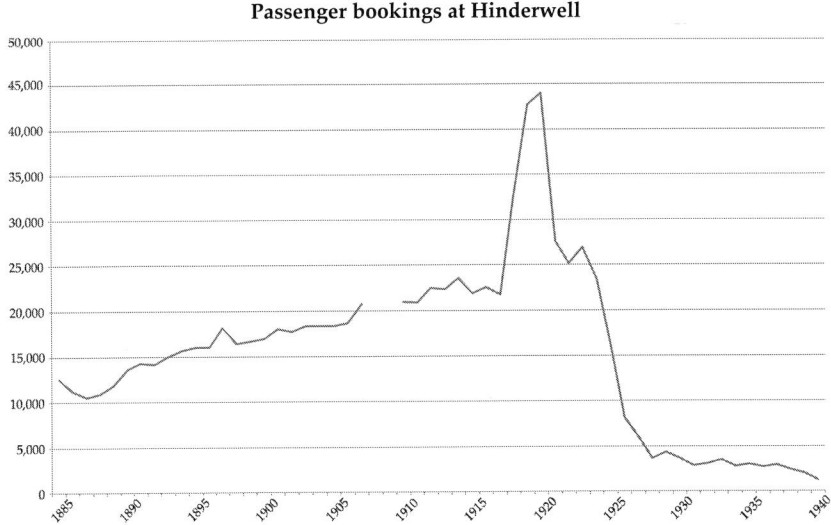

Passenger bookings at Hinderwell

Staithes

Probably of all the stations on the line, the promoters of the WRMUR had the highest hopes for Staithes. The 1880 Prospectus, discussed in an earlier chapter, emphasizes the likely benefits to the fish traffic of the area. The station, however, was at the top of a steep hill which led down to the older part of the village and the harbour. The newer part of the village, a 20th century addition, was close to the main Loftus to Whitby road (*see map page 23*).

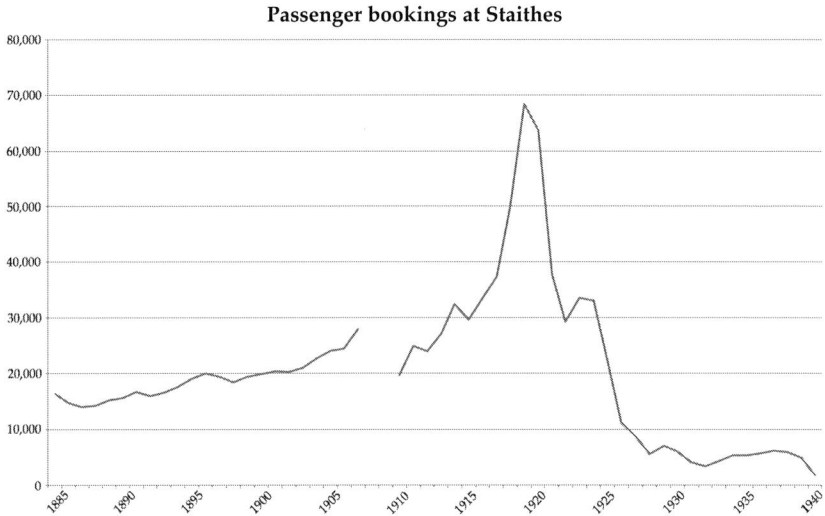

Grinkle

Grinkle station, originally called Easington, was not conveniently placed for the area which it served. Indeed, of all the stations on the line, Grinkle was the most inconveniently placed (*see map page 26*). Easington village was on the main Loftus to Whitby road and consequently as soon as regular bus services were introduced, rail passenger numbers fell away sharply.

Loftus

Loftus station was about two-fifths of a mile from the centre of town and because the road from the town passed through a valley it was not particularly well-placed (*see map page 28*). It possessed large enough sidings for the considerable amount of goods that passed in and out of the station. The annual receipts from the station experienced considerable fluctuations. Also it is likely that the majority of freight traffic made its way to Tees-side, and not along the Whitby line, although there are no figures to support such an assertion. However, as with all stations on the line, Loftus experienced a dramatic decline in passenger numbers booking at the station from the mid-1920s onwards.

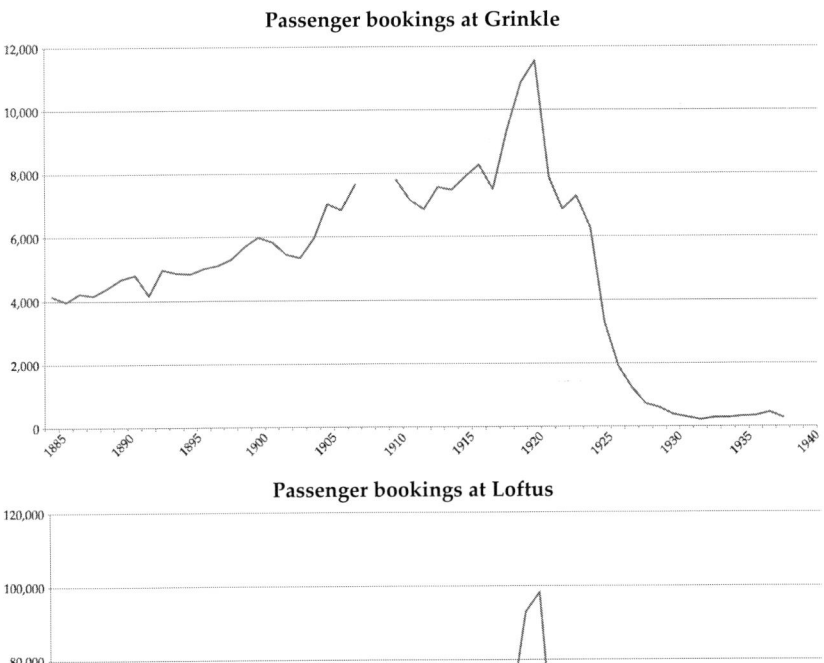

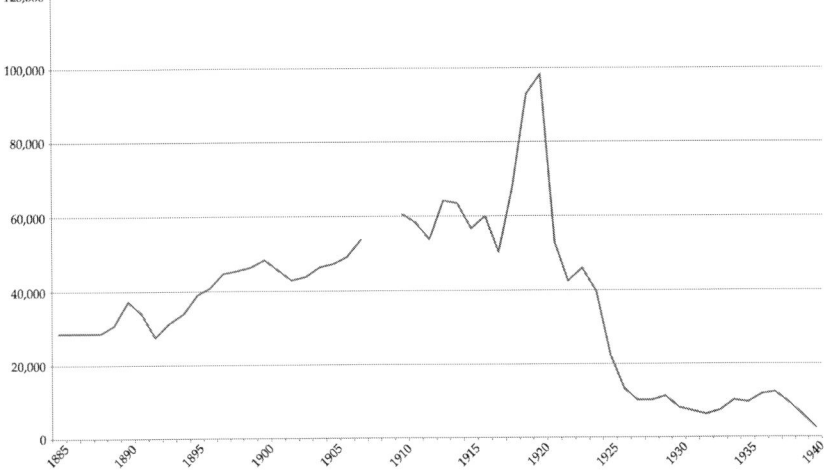

The Whitby-Loftus line suffered badly from the introduction of regular motor omnibus services. Details of these will be given later in this chapter. The proximity of the main road in the villages through which the line passed and the frequency of the bus services posed a threat to the line from which it never recovered.

The Traffic Returns from the Whitby-Loftus line 1897-1907

While by no means spectacular, the total receipts for the line during this period show, with minor fluctuations, a slow but steady increase. These figures, however, are perhaps a little distorted by the inclusion of Loftus, where the goods forwarded and goods received figures show sharp fluctuations in the years 1903-7. However, even with Loftus excluded from the figures for total receipts from stations on the line, there is still evidence of an unspectacular and steady increase.[13] Apart from Loftus, the greater part of each station's annual income almost always came from passenger traffic.[14] For example, in 1897, passenger income at Grinkle accounted for almost 60 per cent of the total receipts; in 1899 at Staithes the figure was over 48 per cent; at Hinderwell in 1901 passenger receipts accounted for over 64 per cent of the total receipts; the figures for Kettleness were even more important, passenger receipts accounting for nearly 87 per cent of the total receipts. At Sandsend, in 1903, almost 63 per cent of the total receipts were provided by the passenger receipts. Finally, at Whitby (West Cliff) in 1905, the total percentage of 85 per cent of the total receipts was accounted for by passenger receipts. The overwhelming importance of passenger traffic to the line is made very clear by these percentages. During the 11 years in question, passenger bookings remained steady, showing growth in the years 1904-7.

Irving's harsh assessment of the Whitby-Loftus line as a 'financial disaster of some magnitude' does, I believe, need to be ameliorated. Irving sampled 596 stations from the NER 1897-1907 traffic returns and, having done so, stated that 30 per cent took less than £1,000 from all sources of traffic, consuming in the process between one-third and three-fifths of revenue in station operating expenses. Nevertheless he maintained that only 20 stations failed to cover their direct operating costs. He does not say which stations these were but, considering the figures for the Whitby-Loftus line, it would seem quite possible that none of these 20 stations were on that line. Irving then makes the assumption that his sample of branch lines as a whole was consuming

74 per cent of their income on operating costs. By 1897 the WRMUR had been completely assimilated by the North Eastern, yet Irving continues to see the line in isolation, giving the gross return from the line as 1.8 per cent from the estimated traffic revenue in 1897 (£11,927) as seen against the final cost of the construction of the line (£655,077).[15] Seeing the line thus, that is, in isolation from the rest of the North Eastern network gives, I suggest, a false impression of the line and seems to contradict his earlier, far more optimistic analysis that only 20 stations of his very large sample could be described as failures.[16] On the other hand, however, the impression of considerable failure which Irving ascribes to the line is emphasized by comparative figures for eight other rural North Eastern branch lines in 1897. Its cost was two and a half times that of, for example, the 23 mile-long Wear Valley line, while its gross return for 1897 was three-fifths that of the two mile-long Easingwold-Alne line.[17]

What did these operating costs consist of? These came from three main departments of the North Eastern Railway: the Locomotive and Traffic department (of which Traffic was the most important); the Permanent Way department; and the Carriage and Wagon section of the Locomotive department.[18] The rise of coal prices after 1890 accounted for another large operating costs, especially in 1900 when coal costs rose by £140,000 alone in one year, with the price of most materials advancing by 20-50 per cent.[19] In the Traffic department, wages and salaries accounted for around 90 per cent of expenditure, while on the railway as a whole they accounted for 26.7 per cent of total expenditure in 1885-89; 27.7 per cent in 1890-94; and 28.56 per cent in 1895-99.[20] The first question posed above, how much did the line contribute to rural mobility, may be answered easily: it probably contributed greatly. However, this conclusion perhaps needs to be qualified by the definition of 'rural mobility'. On the face of it, it means the mobility of those ordinarily resident in the country district through which the railway passes. But there is a problem here, especially with the Whitby-Loftus line; the difficulty is that we cannot assume that bookings from the stations were predominantly made by residents; they could have been by holidaymakers. Leunig made a telling point when he remarked that third class passengers would not, in the absence of railways have travelled by coach, but would instead have walked.[21] Another important point which he stressed was that those who did travel would not have been representative of a cross-section of the working class, for in 1865 workers could only travel 3.3 miles on an hour's earnings; that in 1912 they could travel 10.4 miles on an hour's wages may go some way to explaining the startling rise in passenger numbers on the Whitby-Loftus line from *c.* 1910 onwards.²² Leunig's conclusion and it may certainly be adduced to any discussion concerning the Whitby-Loftus line was that 'people who could never have expected to travel in all their lives were able to do so for the first time. And those who did travel were able to do so more often'.²³

Increased rural mobility, especially for the working class (who we may assume to have used third class to the exclusion of any other), was perhaps the greatest social benefit given by the Whitby-Loftus line. The price of tickets, then, would have had to have been lessened in order to enable the working class to travel by train. Irving gives two main actions taken by the North Eastern Railway which introduced lower pricing: firstly, weekend tickets were issued as day returns (Saturdays only) from 1895, this being extended in 1897 and, secondly, a new range of cheap day, weekly, fortnightly and holiday tickets were brought into existence. Thus in the first decade of the 20th century there was an expansion across the north-eastern region of travel at cheap rates and a growth of the use of return tickets (which were almost always cheaper than twice the single fare); indeed, 60.4 per cent of journeys were made thus in 1900, and 67 per cent of journeys in 1911.²⁴ This, however, could prove to have more to do with enhanced urban than rural mobility.

Two further questions arise from a study of the traffic returns for these years. Firstly, were the timetables adequate for the amount of travelling being done on the line and could an increase in the number, speed and price of trains have improved the early financial fortunes of the line? Secondly, how far did the increase in wages affect the annual total receipts from each station on the line? Firstly, the timetabling: Before the final, complete takeover of the line by the North Eastern Company in 1889, the directors of the WRMUR had complained about the inadequacy of the timetabling, their argument being that the paucity, speed, and price of trains were the main cause of the less than anticipated takings. In early 1886 the WRMUR demanded on the NER an improvement in services because 'the convenience of the travelling public is not as it should be'.²⁵ The NER countered this argument by saying that in the last six months of 1885 passengers booked from all stations between Loftus and Sandsend had decreased.²⁶ It is possible, then, that operating issues could have had an effect upon the line's finances.

For later years, two timetables (from April 1910 and July 1922) are used here. The 1910 timetable will be used to illustrate the findings from the set of traffic returns from 1897 to 1910. The 1910 shows a slight improvement in frequency over the 1887 timetable.²⁷ It should be noted that the service ran from Saltburn to Scarborough; it was not until 1933

A BRIEF FINANCIAL HISTORY OF THE LINE FROM 1897 TO 1940   105

that the northern end of the service was moved to Middlesbrough, with results which will be discussed later in this chapter. A copy of the 1910 timetable is given below.[28] Was this service adequate? The example of Hinderwell provides an answer.

The example of Hinderwell

Hinderwell station was a typical rural branch-line station on the coast of north-east Yorkshire. Its fortunes reflect those of the line as a whole during the period 1897-1940. In 1910 (these figures are from the 1910-34 traffic returns from the line, to be discussed fully shortly) the number of passengers booking from Hinderwell station was 21,024.[29] Working on the principle that the timetable was operated for 317 days in the year (that is, not Sundays), this would mean that an average of 66-7 passengers booked

Public timetable for the Saltburn-Scarborough service (which included the Whitby-Loftus line) for April 1910.   *Bradshaw's April 1910 Railway Guide, Newton Abbot, 1968* at Hinderwell on any given day. There were, in April of 1910, 12 trains a day (6 each way) which stopped at Hinderwell. Of these, four terminated at Whitby (Town), two went all the way to Scarborough while, in the other direction, all six went to Saltburn. On average, then, about six passengers would take each train from Hinderwell.

The service, then, at this time, seems perfectly adequate for the wants of the people of Hinderwell. Do these figures mean the line and the station was 'marginal and unprofitable'? Clearly these passenger bookings are small and could very well be adduced to Irving's argument. Nevertheless, the line was about to enter its (admittedly very short-lived) period of maximum profitability. However, it would be wrong to consider the passenger bookings in isolation, for these do not represent the total use made of a station. Although there are no figures for 1910, in all the years for which there are figures (1913-34) the numbers of tickets collected local to the station far exceeded those who booked there. There was considerable parcel traffic: 6,094 being either forwarded or received that year. The other main sources of income for the station were mineral traffic (1,971 tons of coal, coke, lime, limestone, and ironstone being received) and other goods received (2,126 tons). This receiving of heavy freight far exceeded its forwarding (only 312 tons). Thus Hinderwell was a busy working station whose total receipts that year came to £1,548, of which the passenger receipts came to £1,064, clearly the most lucrative element in the station's generated income. The working of a country station like Hinderwell, though, was very labour intensive, as a photograph taken of the staff at the nearby (and much smaller) Grinkle

Grinkle station in the 1930s *Author's collection*

station in 1905 indicates. Here, at Grinkle in 1905, the total receipts for the year were £569, while the station expenses (wages and salaries) were £176. The station expenses for Hinderwell in 1905 were £197.[30]

What the traffic receipts do not make clear, unfortunately, is whether or not the station expenses (salaries and wages) have been deducted from the figures for total receipts. In 1910 the Hinderwell station expenses came to £227. During the period 1897-1907 the expenses for each year had been as follows:

1897	1898	1899	1900	1901	1902	1903	1904	1905	1906	1907
£253	£263	£250	£250	£257	£202	£207	£207	£197	£202	£204

Hinderwell station expenses 1897-1907.

If the station expenses have to be deducted from the total receipts, this would leave the net receipts as being £1,321. The North Eastern Railway, by the turn of the century, was becoming extremely concerned about the problem of the expense of wages and salaries. There is a clear decline in wages costs before 1901, but after that they began to advance again. Managerial reforms and poor trade between 1901 and 1906 were, Irving suggests, the main reasons for lowering the costs of wages and salaries, but then the percentage of gross revenue taken up by wages increased from 26.8 per cent in 1907 to 28.3 per cent in 1912, thus taking away a large part of the gross revenue up almost 3 per cent in the years 1906-13.[31] However, the proportion of wages to gross revenue was still substantially lower in 1912 than that recorded in 1900-1.[32]

Leunig's conclusion, then, that 'as the period progressed railways offered poor returns to investors, but they delivered tremendous welfare gains to travellers and to society',[33] may be applied to the Whitby-Loftus line. However, there seems now to be a quite severe dichotomy between Leunig's social welfare view and Irving's economic standpoint. Up to the beginning of the 20th century, certainly, Irving's view of the line as 'a more spectacular example of a loss-making branch would be hard to find' does not seem unreasonable in strict financial terms. Indeed, in the context of the North Eastern Railway as a whole 'we should be fairly close to the mark if we assumed that in 1897 our sample of branches (which included the Whitby-Loftus line) as a whole were consuming 74 per cent of their income in operating costs when the parent's (NER) was 58 per cent'.[34] However, the line probably made a contribution to overheads, and this would have been enough for contemporary managers. As S. Joy argued with regard to a later period, 'as long as the revenue for all these stopping services covered their total movement costs, it was thought that they were making a profitable contribution to the railways, because freight and express passenger services were covering all the other costs'.[35] Joy was probably right in saying that this was the stance taken by most railway managers up to the 1950s. What he did not say was that this was an adequate way of costing such services. The Whitby-Loftus line probably covered its operating (that is, total movement costs) from the mid-1890s up to the mid-1920s, but Joy would not have recognized the line as profitable under any criteria he would have used. He would have argued, for example, that stopping services incurred overhead costs that far exceeded the surplus over and above their movement costs. Irving's average and his assumption that the line made a surplus to contribute to overhead costs may be tested by the following analysis of the traffic returns for the line between 1910 and 1940.

NER and LNER traffic returns for the Whitby – Loftus line 1910-1940

We are fortunate to have such detailed figures for the line. The figures for 1910-34 are handwritten in a huge, hard-backed ledger deposited in The National Archives and cover the entire north-eastern region. These returns tell a dramatic story. As illustrated by the graphs shown on pages 97-101, passenger traffic increased dramatically until the high point in 1920, and fell away catastrophically until by *c.* 1926 it was apparent that a very serious decline was occurring; in fact, this decline continued (with minor fluctuations) up until the end of the available evidence in 1940.[36] When looking at the line's annual receipts as a whole, Loftus has been omitted for there are major anomalies in the figures for goods received and sent in some years. Even so, the passenger figures for Loftus are included in this analysis as are the fortunes of all the other stations on the line, for they do not demonstrate the same anomalies as those for freight.

The evidence from the United bus timetables for (some of) the years after 1926 provide the clearest possible explanation for the decline of the Whitby-Loftus railway line. This catastrophic decline in passenger numbers from the mid-1920s onwards was probably nothing special. 'It would seem', wrote Aldcroft of the fortunes of Britain's railways as a whole, 'that passenger traffic reached a peak in 1920, fell sharply and then remained fairly stable though with periodic fluctuations'.[37]

When the question 'how far did the line contribute to rural mobility?' is asked concerning the years 1910-25 then the answer is, considerably. Passenger receipts for those years also suggest that the line, far from being a financial disaster, was doing well. Better, in fact, than it ever had done, and better than it ever would again. However, this was in terms of attracting custom. Financially, though, when the greatly increased operating costs are considered, then this picture is not so rosy. The traffic returns all indicate a steady growth in receipts (with occasional minor fluctuations) to 1918 and then a surge in the years 1919-20. This surge was enormous. It is not easy to explain. It is possible that immediately after the war, there was little else, apart from travel, for the British public to spend on leisure.[38] The summer months of 1919 were warm and sunny, yet those of 1920 were relatively cool and wet, indicating that climate was not necessarily an important factor in the surge.[39] This passenger traffic on the line, for a few years at least, contradicts Irving's assessment that 'passenger traffic ceased to display the vitality of the 1880s and 1890s'. The surge of 1919-20 makes the following rapid decline in numbers even more dramatic. What may now be analysed is the line seen in the context of bus services.

The example of Sandsend

Sandsend station presents a spectacular, but by no means exceptional, example of the changes in passenger numbers, goods forwarded, and total annual receipts which occurred during the 1885-1940 period. There are two major factors which caused this change. Firstly, the building in 1925 of a new, straight, and fast road between Eastrow (*see map, page 14*) and Whitby, thus eliminating the old, winding, steep, tortuous and very slow toll road which had existed beforehand (*see map, page 8*), and the concomitant introduction of a regular bus service between Whitby and Sandsend, as well as the introduction of further bus services, calling at Sandsend, from Middlesbrough, Redcar, and Saltburn. The bus service to Whitby ran through the village, providing a convenience which the railway station (placed someway up a hill above the western part of the village) could not match. The graph of passenger bookings over the 55 years tells a very clear story.[40] Between 1885 and 1908 annual bookings ranged between 22,000-35,000. Then, in the four years before the outbreak of the First World War, passenger bookings were consistently above 42,000 per annum. During the war years bookings fell away to a certain extent, but never fell below 30,000. Then came the two boom years: in 1919 the figure for passenger bookings at the little station was 50,172, whereas in 1920 this figure (along with all other stations on the line) reached its apogee of 54,043. After a fall to 38,883 in 1921 the figures again rise above 42,000 for the next two years before their dramatic fall, which sees an almost complete transfer of passengers from rail to road by 1934 (the low point being 1932, when only 2,264 bookings were made). The numbers of passengers arriving at Sandsend fell dramatically, too, indicating that perhaps a large percentage came from Whitby and, after the opening of the new, toll-free road in 1925, preferred the convenience of the bus. Photographs showing the old road indicate that travel from Whitby to Sandsend by rail was far quicker and probably cheaper. In 1922, for example, 44,819 tickets were collected from passengers arriving at the station (compared with 43,455 booking from Sandsend). Only 10 years later the comparable figures were 3,504 arriving and 2,264 booking.

Public timetable for the Saltburn–Scarborough service (including the Whitby-Loftus line) for July 1922. *Bradshaw's July 1922 Railway Guide, Newton Abbot, 1985*

Taking the same year, 1922, for further examples, parcel traffic was fairly light, with 815 forwarded and 1,874 received. As regards heavier traffic, 1,722 tons were received and 465 tons forwarded. Parcel traffic held up a decade later but the figures for heavier goods were lower, only 33 tons being forwarded and 184 tons received.[41]

Apart from Kettleness (remote, and some way from the main Loftus-Whitby road; the station being more convenient) the dramatic fall in passenger bookings on the line was universal. The July 1922 timetable (which, being at the beginning of the summer season would have catered for a fair percentage of the annual number of 219,898 people that used the line that year if we include Loftus and West Cliff) is not ungenerous in its provisions: 16 trains a day called at Sandsend that summer, five making their way to Scarborough, three to Whitby (Town) in one direction, while six trains made their way to Saltburn in the other direction. There was also a 'local' service of a morning and mid-day train from Whitby (Town) to Hinderwell and back. After Loftus and Whitby (West Cliff) whose bookings no doubt included destinations away from the Whitby-Loftus line, Sandsend was the best-used station in 1922. This was almost certainly because of its position relative to Whitby in that it was situated at the end of the three-mile stretch of beach. A walk to Sandsend along the beach (or vice-versa) and a train back home would provide a pleasant outing or, with children, a return to Sandsend could offer a far less crowded beach than that at the Whitby end. Nevertheless one would have to time one's journey carefully, something that was not necessary with the bus service which seems to have begun with the opening of the new road to Whitby. In 1929 (the first Whitby-Sandsend timetable available) there were 28 buses (running more or less every half hour) during the day and evening:

Omnibus Company Service 13 (Whitby-Sandsend), 1929.
The Omnibus Society library, 100-2 Sandwell Street, Walsall (no catalogue reference)

The Sunday bus service, while less comprehensive than the weekday service, was nevertheless far superior to that of the railway, which ran no trains at all through Sandsend on Sundays (in 1922).

The United timetables also indicate that not only was Service 13 available for visitors and residents of Sandsend, Service 35, which in April 1926 (the earliest timetable available) ran from Middlesbrough as far as Hinderwell, offering six through buses on weekdays, with four through buses on Sundays (in 1926 the passenger bookings at Hinderwell station reached a new low of 8,210; the high point had been in 1920, when 44,044 bookings were made. By 1931 the figure had fallen to 2,818). By October 1927 (the next timetable available), the 35 service had been extended to Whitby. There was an hourly service from Middlesbrough to Whitby beginning at 7.50 am and then from 1.50 pm a half-hourly service until 6.50 pm. The last through bus left Middlesbrough at 8.50 pm. The return service offered regular through buses from Whitby to Middlesbrough, beginning at 9.20 am This ran hourly until 2.20 pm when the service was half-hourly until 7.20 pm, with the last service leaving Whitby at 8.20 pm. Certain late buses ran as far as Skelton. As far as Sandsend was concerned, this meant that in October (beyond the holiday season) the 35 service provided 19 buses in each direction (with a Saturdays only late bus to Loftus). This service called at every village along the line, with the exception of Kettleness. As the steepest falls in passenger bookings occurred from 1925 onwards, it seems undeniable that the frequency and convenience of the new bus services were the cause of that fall. In 1928 the 35 service was extended to Scarborough. However, the bus took almost four hours to make the journey (3 hours and 55 minutes) in 1928, the train, in 1933 (when the northern terminus changed from Saltburn to Middlesbrough), took between 2 hours and 25 minutes, and 2 hours and 40 minutes (according to whether the train stopped at all stations).[42]

It would seem likely, then, that shorter journeys were transferred from rail to road, while the longer journeys, for example from Middlesbrough to Whitby and from Middlesbrough to Scarborough, were undertaken by rail. Then, in 1928, a further service was added: that from Redcar to Whitby (Service 34). Again, this passed through all the villages on the line. For the timetable for the dates 1st July to 18th Sept (that is, the height of the holiday season), 26 buses passed through Sandsend on their way to Whitby, while 23 (one to Loftus only) passed through the village of their way to Redcar. Astonishingly, in 1929, during the height of the holiday season (8th July-17th September) 114 buses passed through Sandsend from Whitby. These comprised the Whitby-Sandsend local (service 13: 48 buses), Whitby-Redcar (service 34: 24 buses), Scarborough-Middlesbrough (service 35: 31 buses) and Whitby-Saltburn (service 25: 11 buses). The railway simply could not compete with such a service, which explains exactly why the passenger numbers fell so catastrophically. The only village through which the main road did not run and which buses did not serve, Kettleness, suffered less than any other station on the line, as may be seen from its graph for passenger bookings 1885-1938.[43]

Grinkle station, which was the most inconveniently placed of all stations on the line, suffered so greatly from the coming of regular bus services that, by 1932, the passenger bookings had fallen to as low as 228. The high point, as with all the other stations, had come in 1920 when 11,552 passengers had booked tickets at the station. The graph for passengers booking at Grinkle in the years 1910-38 reflects the fortunes of all the stations (except perhaps Kettleness).[44] By 1928 Grinkle's station expenses (£512) exceeded that of the total income for the year (£508).

Is Your Journey Really Necessary?

The number of passenger bookings on the line lessened even further with the outbreak of war in 1939. The figures for the first year proper of the war (1940) indicate the steepest drop in numbers travelling since the disastrous years of the mid-1920s. This may very well have been due to wartime propaganda concerning travel, the purpose of which was to ensure – as far as was possible – that transportation of servicemen, war supplies, and munitions (and, in 1939 and 1940, evacuees) had to take priority over passenger journeys.[45] That this propaganda had some effect locally is clear from the following figures showing passenger bookings on the line in 1938, 1939 and 1940:[46]

Station	1938	1939	1940
Loftus	9,621	6,303	2,633
Staithes*	6,020	5,065	1,824
Hinderwell	2,396	2,016	1,249
Kettleness	3,367	2,839	2,032
Sandsend	2,687	2,143	880
West Cliff	19,655	16,775	8,018

* The figures for Grinkle station were so low (290 in 1938) that it was temporarily closed in 1939. What passengers there were for 1939 are included in the Staithes figure. The station never re-opened.

Kettleness station showed the lowest fall in bookings, indicating that it still provided a necessary (rather than optional) service to that relatively remote area. The greatest fall numerically was at West Cliff, but the station showing the lowest number of bookings in 1940, Sandsend, indicated that approximately only three people daily took trains from that station. This was consistent with the sharp decline of the 1920s, where Sandsend showed the most dramatic fall in usage.

The traffic receipts for both periods, then, enable the questions posed at the beginning of the chapter to be answered. Firstly, how far did the line contribute to rural mobility? The answer must be: considerably, up to $c.$ 1926, then less and less until, by 1934 its contribution was minimal. Secondly, can Irving's two statements ('the line was a financial disaster of some magnitude' and 'passenger traffic ceased to display the vitality characteristic of the 1880s and 1890s') be justified by the history of the line between 1897 and 1934. Again, the answer – on the whole – must be no to both statements. Irving's first statement dealt mainly with the construction costs of the line and its early income, but this income improved considerably up until $c.$ 1926 and, too, the passenger figures indicate that the passenger traffic (again, up to $c.$ 1926) was certainly displaying a 'virility'. Should the line have been built at all? The answer here must be equivocal, for while, costs were high and the profits low (if non-existent by the late 1920s), it provided a most important social service to the inhabitants of the villages through which its passed, as well as providing (as the figures show) an important, indeed, the only, method of transporting freight. But perhaps most importantly, and this becomes very clear indeed after 1933, the line must be seen, not in isolation, but part of the wider network; while passenger bookings at the village stations fell massively, through trains from Middlesbrough to the coastal resorts of Whitby and Scarborough were (until the advent of popular motor car travel from the mid-late 1950s onwards) extremely well patronized, at least in the summer. It was this factor that enabled the line to survive (at least after 1933) the catastrophic fall in passenger bookings. The line's existence, certainly in its first 50 years of existence, was undeniably justified by its social function rather than any other element.

CHAPTER SIX

THE IMPORTANCE OF FIELDWORK IN RESEARCHING THE HISTORY OF THE WHITBY – LOFTUS LINE

Sources for the history of a railway line are of three kinds: written, visual (that is, film and photographs), and material (that is, the physical remnants of the line itself). It is the latter element which is emphasized and commented upon in this chapter, offering some examples of how fieldwork (that is, visiting the physical remnants of a line) can make a considerable difference to a railway historian's research.

Such fieldwork will often clarify and even improve upon the information given in the primary sources which may exist concerning a line's history. For full and proper research the remains of the line should be visited as many times as possible, for only an intimate knowledge of the line's topography and the remaining visual evidence of construction can offer the historian a deeper understanding of the problems involved in that line's construction, especially if the evidence used so far in research (while no doubt excellent) is only of the written or photographic kind. For the railway historian the Whitby – Loftus line offers a wealth of written primary source material, most of which is deposited at The National Archives in Kew.

The historiography of the Whitby – Loftus line

Until 2010, very little had been written about the line. Certainly there had been no academic study of its history. Indeed, before its end in May 1958 its history consisted of five articles (two very specialist).[1] A further article was written in the August of that year. The main - indeed the only – text devoted entirely to the line is Ken Hoole's 1981 publication, which has its origins in a chapter in his 1971 work on Cleveland railways.[2] There have also been a number of books, mainly published in the last 25 years, which deal with the railways of the region, and which give some reference to the line. The photographic content of these texts is useful, but little new is said (apart from the occasional personal reminiscence) and what is said is mainly based on a reading of Hoole.[3] There are, however, two very important visual sources. The first was made on a cine camera by Arthur 'Cam' Camwell and shows a trip on the line between Upgang viaduct

(Whitby) and Loftus.[4] There are also sequences showing Staithes viaduct and Hinderwell and Sandsend stations. The film was made in either 1956 or 1957.[5] The second item, made in probably 2007 and 2008 is a 10 minute film of the interior of the abandoned Deepgrove (Sandsend) tunnel. This is of considerable interest as it shows the quality of the workmanship of the tunnel, the spoil tunnels leading from the line to the cliff face, and the ventilation shafts Although the tunnel has been abandoned for 50 years, it remains in remarkably good condition (except for a landslip at the western (Kettleness) end)[6] and emphasizes the demands for a line of the highest standards made by the three inspections of the Board of Trade before the opening of the line.[7] A viewing of this film makes clear that not only was this line very difficult to construct, but that it was not done cheaply. Although in this instance the fieldwork (exploring the abandoned tunnel) was done not by railway historians but by explorers of such artefacts the video illustrates the main argument of this article.

The most important secondary source is Hoole's book on the line. It is evident that he had consulted most – but by no means all – of the primary sources. Most later writing on the line owes a considerable debt to Hoole, while it is clear that later writers have not consulted the primary sources, but merely accepted Hoole at face value. This is all very well for the casual reader, but by no stretch of the imagination can any other text apart from Hoole be considered as a reliable source. Hoole's book is, unfortunately, unpaginated, and is mainly of photographic and ephemeral content, with only three pages of text which deal with the origins of the line and a section on 'Stations and Trains', which concerns itself with the working on the line and a brief description of the stations thereon. Hoole, the doyen of north-eastern railway history, wrote widely about many aspects of that history and, naturally, could not be expected to visit every line and study it *in situ*. Nevertheless he did visit the Whitby-Loftus line upon occasion and his photographs of the line are invaluable.

However, since 2010, the line has been the subject of a York University M.A. thesis[8], a book[9], and four articles in *The Journal of the Railway and Canal Society*.[10] Since the completion of the thesis, the publication of the book and the first three articles, the author has returned to live in the immediate area. Visiting the remains of the line on numerous occasions, particularly between Sandsend and Kettleness, he has become aware that, while the primary sources retain their enormous importance, much remains hidden or obscure in those sources which can only be revealed and understood through regular

Remains of alum working at Sandsend showing the permanent disfigurement of the cliffs.
Author

fieldwork. There are a number of examples which may be given to illustrate this argument, three of which follow.

Steeping Pits viaduct

To the west of Sandsend, the line was constructed, mostly on embankments, over disused alum workings. These workings which operated for over 250 years (1605-1871) have badly disfigured the cliffs. The production process was long and complicated; it involved extracting then burning huge piles of shale for at least nine months before transferring it to leaching pits to extract an aluminium sulphate liquor. This was then sent to the steeping pits where human urine was added. The Sandsend steeping pits are clearly visible from the railway embankment to their immediate north. (*For map details, see page 14.*)

The Directors' minutes of the Whitby, Redcar and Middlesbrough Union Railway sometimes (but by no means always) include a written copy of the most recent Engineer's Certificates. In 1873 there are five references to the Steeping Pits viaduct in these reports (*see pages 119 and 120*). They appear to be cumulative (that is, they show the total expenditure so far upon a

Engineer's Certificate:— A certificate No 21 for works &c was submitted by Mr Hamond on behalf of Mr Filone as follows viz:—

Earthworks		6,009.3.4
Masonry		172.10.0
Culvert (6ft Culvert 30 Lgth)		251.0.0
Bridge No 1		80.0.0
Sleeping Pit Viaduct		221.0.0
Iron Viaducts		3,500.0.0
Tunnel		1000.0.0
Permanent Way (Rails)		2,500.0.0
		13,611.13.4
Cleveland opposition Solicitors	229.2.6	
Engineer	245.7.3	474.9.9
Solicitors		595.9.5
Engineering		400.0.0
Secretary		163.4.6
Tenants Compensation (Land)		370.15.8
		1,998.6.10
		15,609.19.7
Deduct Certificate No 20		1,1263.12.3

Copy of Engineer's Certificate for 4th April, 1873.

Copy of Engineer's certificate for 11th July, 1873.

Engineer's Certificate:— The Engineer submitted a Certificate No 24 for Works and Materials as follows:—

Earthworks		11,212.7.6
Masonry		916.15.0
6 feet Culvert		250.0.0
Bridge No 1		80.0.0
Sleeping Pit Viaduct		200.0.0
Iron Viaducts		5,500.0.0
Tunnels		2,266.0.0
Laying Roads		100.0.0
Ballast		250.0.0
Rails (400 tons)		3,200.0.0
		23,974.2.6
Cleveland opposition		474.9.9
Engineering		800.0.0
Solicitors		791.1.9
Directors		50.16.7
Secretary		217.9.7
Auditors		20.0.0
Land		374.15.8
		2,728.13.4
		26,702.15.10
Less Certificate No 23		23,653.4.6
		3,049.11.4

Engineer's Certificate. The Engineer submitted a Certificate, No. 25, for Works
Materials &c. as follows:—

	£ s d
Earthwork	13,972 — —
Masonry	1,058 — —
6 ft. Culvert	250 — —
4 ft. do	270 — —
Bridge No. 1	80 — —
Steeping Pit Viaduct	200 — —
Iron Viaduct	7,500 — —
Forwards —	£23,330 — —

Copy of Engineer's Certificate for 22nd August, 1873.

Copy of Engineer's Certificate for 20th October, 1873. Note that the amount of expenditure on the Steeping Pits viaduct has increased by 33 per cent.

125 20 Oct. 1873
Engineer's Certificate No. 27. The Engineer submitted a Certificate, No. 27, for Works, Materials &c. as follows:—

	£ s d	
Earthwork	18,448 6 8	
Masonry	1,740 — —	
6 ft. Culvert	250 — —	
4 ft. do	300 — —	
Bridge No. 1	80 — —	
Steeping Pit Viaduct	300 — —	
Iron Viaducts	7,500 — —	
Tunnels	4,681 — —	
Laying Road	450 — —	
Ballasting	350 — —	
Retaining Wall	333 15 —	
Fencing	918 2 —	
New Road	150 — —	
Diversion of Old River course	194 17 6	
Rails	6,840 — —	
		42,536 1 2
✓ Cleveland Opposition	774 9 9	
✓ Engineering	1,600 — —	
Solicitors	963 8 6	
Directors	61 13 1	
Secretary	306 6 9	
Auditors	20 — —	
Land and Tenants' Compensation	624 15 8	
		4,350 13 9
		46,886 14 11
Deduct Certificate No. 26		40,222 12 0
		6,664 2 11

And the same was...

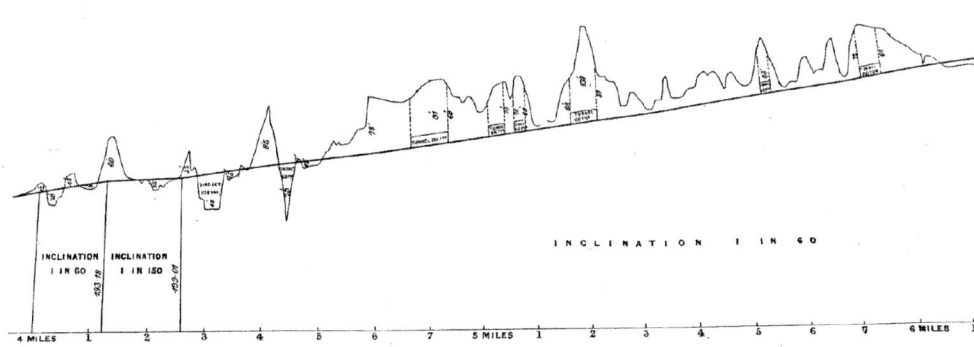

1873 plan showing the two viaducts (Steeping Pits and Overdale Beck) and the six tunnels which were to be constructed around the edge of the cliffs between Sandsend and Kettleness. The distance in miles is from Bog Hall Junction, Whitby.

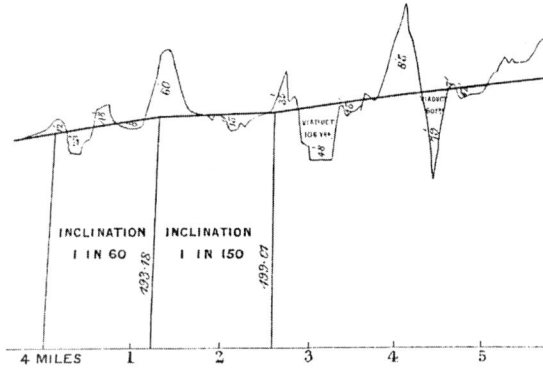

Close up of the plan above showing both viaducts.

particular project). Work was continuing on the Steeping Pits viaduct throughout the year, the final expenditure being £300.[11] It will be noticed that expenditure of the construction of John Dixon's iron viaducts is written as a separate item.[12] Given the difference in cost of these two items, it may be reasonable to consider the possibility that the Steeping Pits viaduct was constructed of wood. Also, interestingly, the Engineer's certificates mention expenditure on tunnels. These were the tunnels which were constructed along the original line which ran around the cliffs.

The 1873 plans indicate that two viaducts (at Steeping Pits and Overdale Beck) as well as six tunnels were to be constructed around the cliff edge between Sandsend and Kettleness. Steeping Pits viaduct was to be 106 yards (97 metres) in length.[13]

However, without the necessary fieldwork, it would be difficult for the railway historian to realize that not only have all six tunnels vanished but that the Steeping Pits viaduct is now an embankment and the Overdale Beck viaduct was never built at all, although the embankment on the far side of the steep beck valley which still remains was clearly constructed for a viaduct to abut onto it. There is no record of the Steeping Pits viaduct after October 1873 (which, had £300 spent upon it and was possibly finished or almost so) in any of the remaining primary sources. It may very well have been demolished, but this would have been a long and rather unnecessary task. It would have been much easier to cover it as happened with Kilton viaduct near Loftus (1911-13) after subsidence had made that original viaduct dangerous.

So why was the Steeping Pits viaduct covered? It may have been dangerous or flimsy, but this was unlikely given the relatively large amount of money that had been spent on it. There was no real need to cover it, but it seems likely that the reason for covering it was because there was nowhere to put the vast amounts of spoil taken from the Deepgrove tunnel excavations, the tunnel being almost a mile in length. The eastern portal of Deepgrove tunnel is a few hundred yards from the Steeping Pits viaduct. We know that Deepgrove tunnel was begun in July 1879 and not completed until 1882. Some of the spoil was taken along two specially constructed adits in the middle of the tunnel and

Steeping Pits embankment from Deepgrove. In the foreground are the remains of the steeping pits. *Author*

Steeping Pits embankment from the seaward side. *Author*

Spoil tip (to the immediate left of the line), with Steeping Pits embankment visible in the middle distance. In the foreground where the line comes into view is the approximate position of the 1927 derailment. *Author*

dumped over the cliffs into the sea, but most had to be taken to the exits. Today, these spoil tips have become part of the landscape, but there is no disguising what they originally were. When the North Eastern Railway took *de facto* control of the line in 1875, a number of demands were made. The main demand was that the cliff edge line be abandoned and a tunnel (the present Deepgrove tunnel) through the cliff built further inland. We also know that the construction of the line under John Dickson was so inept that almost three years had to be spent in remedial work.[14]

Only by regularly walking, exploring and observing the line and its environs (that is, fieldwork) can the argument for the possible fate of the Steeping Pits viaduct be appreciated.

Owen's Cliff Signal Cabin

Of all the bleak, lonely, windswept, desolate, remote and dark places on the railways of England in 1883, none was more so than Owen's Cliff Signal Cabin. Between Deepgrove (a mile west of Sandsend) and Kettleness (*see page 19 for map*) the railway ran in the open for 100 yards or so between two tunnels, one a mile in length and the other over 300 yards long, along a ledge hundreds of feet above the sea, and hundreds of feet below the summit of the cliff above. Signals guarded the

The line between Deepgrove and Kettleness tunnels.
J.W. Armstrong/Armstrong Photographic Trust

entrances to the tunnels and a signal cabin had been built to control these signals. When the line opened, clear instructions were given to the drivers, firemen, guards and signalmen along the line as to the position and importance of these signals. This document is unique, and possibly the rarest of all such sources to come to light concerning the line.[15]

Trains were few and far between and for the signalmen who worked the shifts, which would have been lonely and desolate experiences, especially in the dark winter months. But where is Owen's Cliff? And where was the signal cabin? Detailed maps of the terrain between Kettleness and Sandsend name all the cliffs, plateaus, valleys, scaurs and headlands; it seems that every hill and dale has a name, such as Lucky Dog Point, Overdale, Holmsgrove, Seavybog, Tellgreen, Keldhowe, Stonecliff End and Deepgrove, but the name Owen's Cliff is nowhere to be found on any map published from the 17th century to the present day.

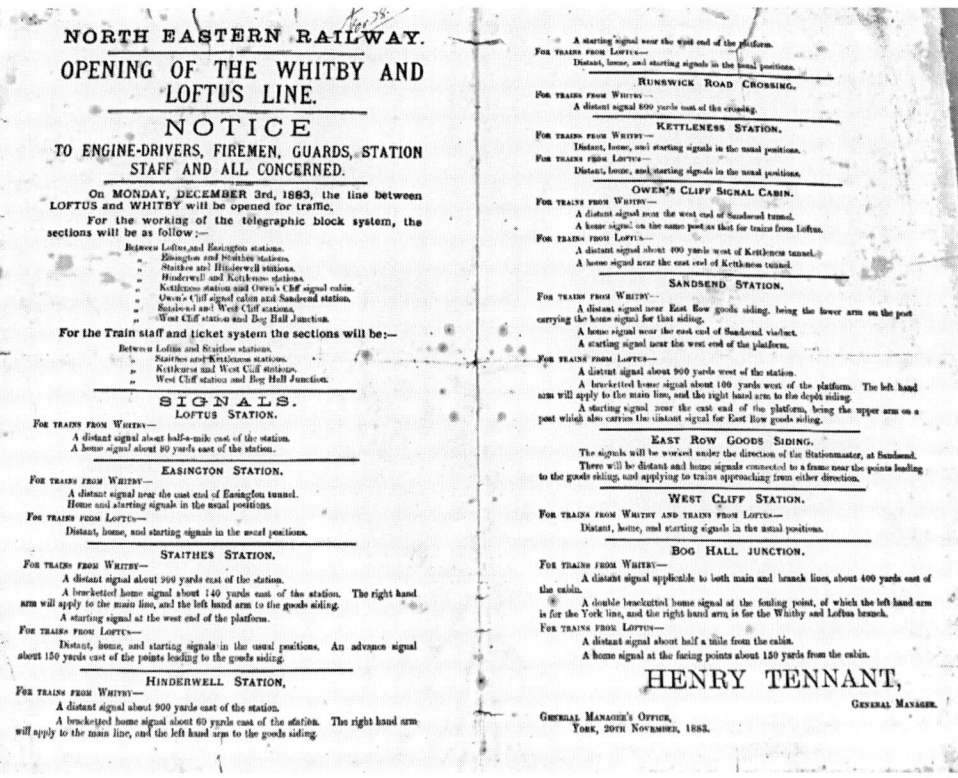

The General Manager of the North Eastern Railway's instructions to drivers, firemen, guards and all concerned with the new line between Loftus and Whitby. *NRM, York*

> **OWEN'S CLIFF SIGNAL CABIN.**
> FOR TRAINS FROM WHITBY—
> A distant signal near the west end of Sandsend tunnel.
> A home signal on the same post as that for trains from Loftus.
> FOR TRAINS FROM LOFTUS—
> A distant signal about 400 yards west of Kettleness tunnel.
> A home signal near the east end of Kettleness tunnel.

Detail from the NER's instructions to drivers showing the positions of signals to both the west and east of the smaller tunnel (Kettleness) and to the west end of the mile long Sandsend (Deepgrove) tunnel.
NRM, York

This cliff named by Henry Tennant in his instructions can only be that which lies above (and below) the railway line (*see photographs page 123 and 126*). But why Owen's Cliff? In the mid-19th century the geologist Professor Richard Owen (1804-92) was, probably, second only in fame to Charles Darwin, although he is almost unknown today. Owen was instrumental in the establishment of the Natural History Museum in London and it was he who coined the term 'dinosaur'. At this time, a number of fossil reptiles – ichthyosaurs and pre-historic crocodiles – were discovered in the Kettleness alum quarries. These finds were discussed in some detail by Owen during the height of his fame. It may very well be that these dinosaurs and their popularization by Professor Owen led to Henry Tennant, not knowing local names, calling the signal cabin Owen's Cliff, perhaps mistaking the area where the line was built from where the dinosaur remains were found.

The whereabouts of the cabin presents a different problem. Apart from Frank Meadow Sutcliffe's famous photograph of Kettleness tunnel under construction in 1881, there are no known photographs of this stretch of the line until the late 1940s. What is known is that the Sandsend signal cabin, mentioned in the Tennant instructions of 1883, was de-commissioned in 1905 and the local signals removed. It is likely that the Owen's Cliff signal cabin was closed then, for no signals existed after that date between Kettleness and Whitby (West Cliff).

It is not difficult to imagine that the weather conditions at Holmsgrove (the correct name for the area) were so adverse that it proved difficult at times to operate the signals. The line there was open to winds, rain and snow from the north and east. The signal cabin would have been placed in the area which provided the greatest shelter. There are two possibilities: the most likely is that the cabin (it must have been very small) was built into the space cut into the almost sheer face of the cliff which rises dramatically above the line. In a 1940s photograph

Sandsend station c. 1905. The signal cabin can be seen at the far end of the platform. Owen's Cliff cabin would have been smaller. *Author's collection*

Holmsgrove near Kettleness. Before the mystery was solved, speculation as to the whereabouts of Owen's Cliff signal cabin suggested that it might have been sited in the space dug into the cliff. *C.C. Cobb*

(*lower facing page*) this space is visible just before the abutment to the east portal of Kettleness tunnel.

The only other possibility is that the cabin was built on the sea side of the line just as it emerged from Deepgrove tunnel. Another photograph from the 1950s shows that at this point there was a certain amount of protection from the elements by the nearby cliffs (*lower page 75*). Today this is the most inaccessible section of the line. Between the two tunnels the area is completely overgrown and would involve a very steep climb down the cliff from above. It is, however, possible to access this part of the line if one walks through the 308 yard long Kettleness tunnel (entering through the western portal). Fieldwork is, indeed, possible and occasional group visits are made and the tunnel entered and explored. The only possible way to walk the trackbed between Kettleness and Deepgrove tunnels is in the depths of winter when the foliage is at its least fecund. Nevertheless this is dangerous and the greatest of care should be taken.

The mystery solved

After writing about the mystery in the *Journal of the Railway and Canal Historical Society*,[16] correspondence in reply to the article ensued. One reply was clear:

> 'Owen's Cliff signal box is shown as 'S.B' on the OS 6 inch map re-surveyed in 1893 and published in 1895. The box itself is even more clearly marked on the OS 25 inch map of 1894 just south of the mouth of Kettleness tunnel but on the *opposite* (coastal) side of the line from the position Dr. Williams speculates. The signal post carrying both home signals is also marked. Both box and signal post have gone by the edition re-surveyed in 1911-13 and published in 1919.'

Correspondents on this issue have suggested two reasons for the positioning of the box, whose active life lasted from 1883 to *c.* 1905: either that the box's function was to break up the block section and to allow two trains running in the same direction to follow each other more closely that would otherwise have been the case, or that the box was placed in that position for safety purposes; given the exposed situation it could give an early warning of landslides and damage caused by gales, for example. The Tay Bridge disaster cannot have been far from people's minds. Once again, it is a combination of diligent research (the Henry Tennant document may be found at The National Railway Museum, York), and determined fieldwork which enables the historian to understand this apparent mystery of Owen's Cliff Signal Cabin.[17]

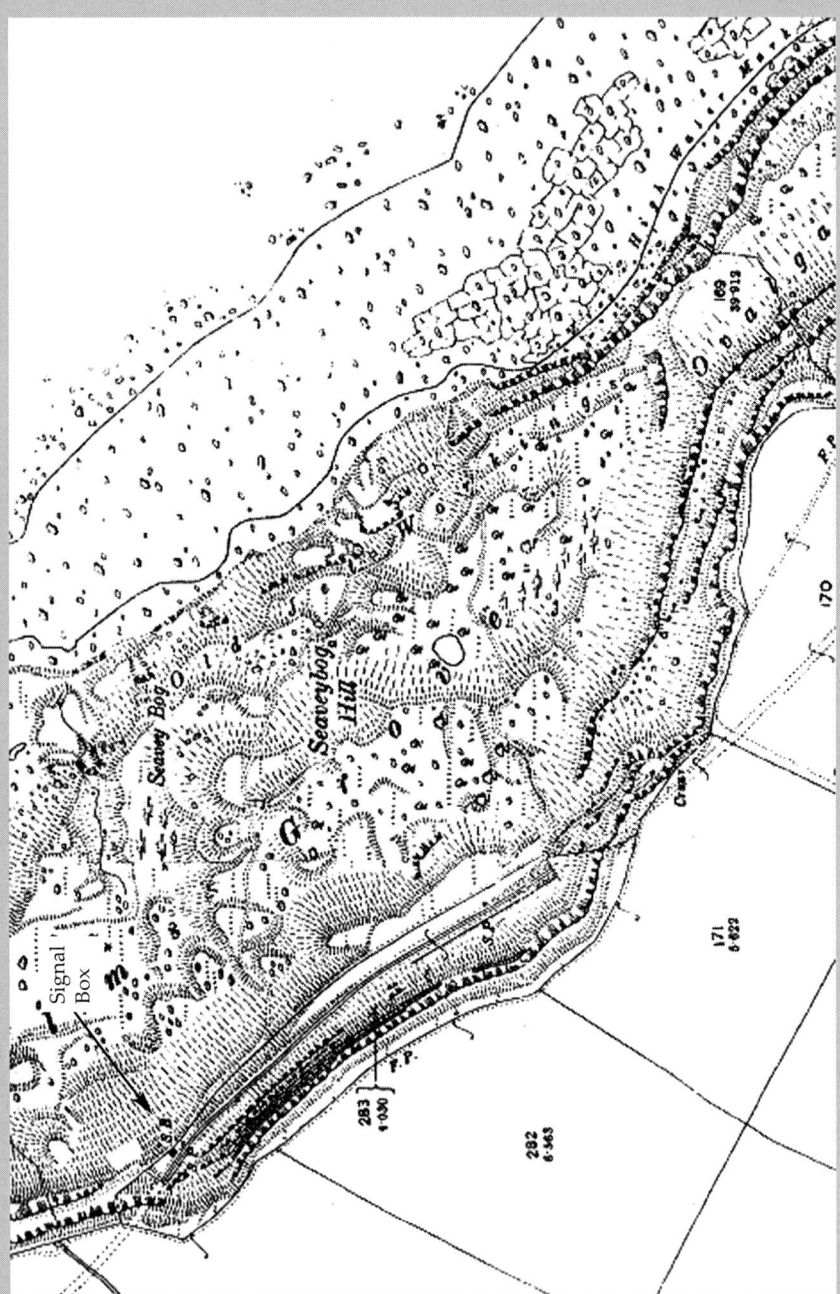

O.S. 25 inch map (1894) showing the 300 yards or so of open line between Kettleness and Deepgrove tunnels.

The derailment of 1927: a very close call

Whitby and the surrounding area has been remarkably free from railway accidents. But there was one near miss which, had it occurred a few seconds later, could have led to a major disaster. On Sunday 31st July, 1927 at about half past nine in the evening a return excursion train from Whitby to Hartlepool passed through Sandsend station travelling at approximately 30 mph. The gradient up to the next station, Kettleness, was a steep one at 1 in 57; it lasted for almost three miles and drivers on the line, if they were not booked to stop at Sandsend as this excursion was not, liked to build up as much speed as possible so they could take a run at the gradient as it was difficult work for the fireman. It had been a lovely, warm and sunny day in Whitby and the returning excursion was packed. It was a very long train, nine carriages, and so it was necessary for the train to be double-headed for the steep gradients and sharp curves between Whitby and Saltburn, where the line to Hartlepool became easier.

About 80 yards from the Deepgrove tunnel mouth. It was here that the derailed train came to rest.
Author

Inside Deepgrove tunnel today. A derailment here at 30 mph would have been disastrous.
Author's collection

As the train approached Deepgrove tunnel at a speed somewhere around 30 miles an hour the lead engine derailed, causing the second, or 'train' engine and six of the nine carriages to derail. 'The derailment appeared to have taken place 200 yards east of (before reaching) the tunnel mouth and to have travelled some 120 yards to a point 80 yards from the tunnel entrance and the permanent way for this portion was very damaged.' This quote is from a report made to the London and North Eastern Railway by the Kettleness station master.

Had this derailment taken place a few seconds later, *inside* Deepgrove tunnel, then there would have been terrible consequences, for the most dangerous of all railway accidents take place in tunnels. The Kettleness station master reported that, luckily, the locomotives and carriages had remained 'head on' and had not slewed across the track. In the narrow confines of the tunnel this is exactly what would have happened and massive destruction and even possibly fire would have occurred. It was because the train remained 'head on' to the track that only two casualties (minor injuries) were reported.

The cause of the accident was straightforward. Because of the length and weight of the excursion train it had been necessary for a more powerful engine to be attached ahead of the 'train' engine. The weight of this engine, plus (possibly) the speed at which the train was travelling caused the track to 'spread' and cause some of the wheels of both engines and six of the carriages to become derailed. It appears that lessons were not learned from this accident as the same problem occurred 10 years later at Prospect Hill Junction (Whitby) when a train to Scarborough was derailed with the same class of engine. Luckily the train was travelling slowly and there were no casualties.[18] After this second accident the entire class of engine (a 'J39') was banned from use in the Whitby district. At Deepgrove, the derailed train was cleared, the track quickly relaid and normal operations resumed at four o'clock in the afternoon of the next day. Perhaps of all the examples given in this chapter regarding the value and importance of fieldwork, this is the one which best illustrates its value. The original document, little known as it is, is transcribed in *Appendix Four*, with the official summary of the incident shown as *Appendix Three*. In itself the incident seems relatively trivial. Indeed, the *Whitby Gazette* relegated the story to its inner pages and the editor decided that the item was so un-newsworthy that it gave only a brief paragraph to a report on the derailment.[19] However, it is immediately apparent to anyone researching the incident by means of fieldwork that a major disaster was averted only by seconds. Less than a minute later and four or five hundred yards further on, the train would have been in the close confines of the tunnel; a derailment at 30 mph does not bear thinking of.

The uses of fieldwork

Of course, not every deserted and abandoned line can be visited. Many have completely disappeared with redevelopment of the land. Neither are the sterling efforts of railway historians researching their primary sources in archives and libraries to be decried. Nevertheless, experience has taught me that if the line under research does have a reasonable abundance of physical remains and if that line is accessible, then regular fieldwork can be of immense assistance to augmenting the writing of a line's history. Historians are like detectives, searching for clues to solve a mystery. All railway historians will understand and will have probably experienced the excitement of finding a document which has seemingly been overlooked in the past. It is the same with fieldwork; often clues hitherto unknown become apparent when manifested visually. Again, speaking from experience, sometimes clues can stare you in the face for weeks, sometimes months before becoming obvious. For example, just beyond Overdale Beck close to Deepgrove tunnel an embankment exists which was part of the original trackbed (*see plan page 120*). Because it had been an accepted fact for so long that no remains of the original line exist, I took this at face value, never thinking that such a trope could be completely incorrect until walking along the track one day I saw (as it were through new eyes) what could only be a railway embankment. And yet I had looked at this embankment many times before, it was just that I did not make the necessary connection. Now a completely new element in the line's history could be explored. This, *inter alia*, is the value of fieldwork.

A Middlesbrough-bound train crosses Eastrow viaduct in August 1957.

Oakwood collection

CHAPTER SEVEN

THE SUEZ SPECIALS

In the 1950s hundreds of men from Whitby and the immediate locality worked at I.C.I. Wilton works, near Redcar. Most of them travelled to work by bus. At the same time the railway service which ran from Whitby via Loftus to Middlesbrough had been reduced to three trains a day each way and was attracting very little custom as the villages through which the line ran were better served by road. It seemed that the line, particularly the Whitby (West Cliff) – Loftus section which ran spectacularly along the coast via Sandsend, Kettleness, Staithes, and Hinderwell was doomed.

But in late 1956 an international incident occurred which, for a brief period of time, altered the fortunes of the line in a most dramatic manner and, if managed correctly, could have saved the line. This incident was the well-known Suez Canal crisis which, in the long term indicated the decline of Britain as an imperial power and, in the short term, produced a fuel crisis owing to the enforced closure of the Suez Canal by the Egyptian leader, Colonel Nasser. The shortage of petrol had a most dramatic effect on the fortunes of the Whitby-Loftus line and showed what an important link the line could be. The story which follows contains information, some of it very detailed indeed, from the pages of The *Whitby Gazette* which gave detailed coverage to every local incident of importance.

Beyond Brotton, the line branched: one line leading to Guisborough and Middlesbrough, while the other (still operational today) made its way to Saltburn West Junction and then to Marske and Redcar. It was this latter branch which is at the centre of our story. By the middle of December 1956 the fuel shortage had begun to bite and it was impossible to find enough fuel to run the workmen's buses from Whitby to I.C.I Wilton. According to the district commercial manager of British Railways 640 men needed transportation and, consequently, two special workmen's trains – the Suez Specials – began their tasks on 11th December. Of these 640 men more than 400 came from Whitby while the rest came from Hinderwell and the surrounding villages. This demand produced, it was claimed at the time, the longest train ever to run over the route. A special timetable was produced for the Whitby train and for the Hinderwell train.

The Whitby Suez Special
Whitby (Town) dep.	5.55 am
West Cliff dep.	6.02 am
Sandsend dep.	6.08 am
Loftus dep.	6.42 am
Redcar arr.	7.25 am

The Hinderwell Suez Special
Hinderwell dep.	6.38 am
Staithes dep.	6.43 am
Loftus dep.	6.56 am
Skinningrove dep.	7.00 am
Brotton dep.	7.11 am
Marske arr.	7.28 am

There are a number of points of interest here. The Whitby special ran non-stop (except for calling at Loftus) from Sandsend to Redcar; Skinningrove station which had nominally closed to all traffic in 1952 was being used; there was only a quarter of an hour between trains. Buses met the workmen at Redcar and Marske and carried them to their destination.

The return special timetable was as follows:

The Whitby Suez Special
Redcar dep.	4.50 pm
Loftus dep.	5.23 pm
Sandsend dep.	6.01 pm
West Cliff dep.	6.08 pm
Whitby (Town) arr.	6.17 pm

The Hinderwell Suez Special
Marske dep.	5.07 pm
Brotton dep.	5.24 pm
Skinningrove dep.	5.32 pm
Loftus dep.	5.37 pm
Staithes dep.	5.53 pm
Hinderwell arr.	6.00 pm

Apparently these trains were popular with the workmen. They were given a type of season ticket (in order to avoid booking each day) and the train times were speeded up, although the district commercial

Passenger trains during the period of 'The Suez Specials' on a snowy day at Loftus. This kind of difficult working would have been encountered daily by the train crews.

J.W. Armstrong/Armstrong Photographic Trust

manager commented that the route was extremely hilly, there was a great deal of single line working, and the rails were slippery in wet weather, especially in the tunnels. A month into the new workings the manager reported that so long were the trains that they over-ran the passing loops in some places; a situation which can only have led to delays.

Yet by the end of February the Suez Specials had stopped, the last running on 23rd February, 1957. The main problem was that the workmen found it impossible to do any overtime as the trains had to adhere to a strict schedule. The petrol situation was improving too, but anyone who has experienced a fairly lengthy bus trip in the 1950s would choose to make the same journey by rail. There may very well be men alive today in Whitby who travelled by these Suez Specials. Thus British Railways lost a wonderful opportunity not only to improve the disastrous finances of the line, but to ensure its survival, for if these trains could run all the year round then they would provide a massive surge in income. However, what the public – and indeed anyone else – did not know was that the decision to close the line had already been made. This decision was not made public until September 1957, but there is written evidence to show that the secret decision for closure had been made by the middle of 1956.

It is interesting to speculate how many carriages would have been necessary for the daily transportation of at least 400 men. Even in the height of summer only five carriages were used on the packed holiday trains from Middlesbrough. I have never seen a photograph of a longer train on the line and I wondered if there was a weight restriction on Staithes viaduct. Looking at photographs of the non-corridor carriages usually used on the line and given that each man would require a seat for the 90 minute journey, I would calculate that the eight compartments in these carriages would each hold eight, perhaps ten men, meaning one carriage might (for argument's sake) contain 70 men. There would then have to be at least eight – and probably nine or ten – carriages to the Whitby Suez Special and at least five for the Hinderwell special. The district commercial manager's rather dramatic description that the Whitby special was the longest train ever to run on the line is somewhat of an hyperbole as we know that in the 1920s excursion trains often contained eight or nine carriages.

But what was the motive power? Double-heading (a train drawn by two locomotives) was possible, but after the derailment of 1927 which

An almost new Standard class locomotive leaves Sandsend with a Middlesbrough-bound train. This kind of locomotive would have been ideal for the lengthy and heavy Suez Specials. *J.W. Armstrong/Armstrong Photographic Trust*

A Middlesbrough-bound train about to enter Deepgrove tunnel in summertime. This was the route used by the 'Suez Specials'. This is the site of the 1927 derailment which, had it happened in the tunnel, would have been a major disaster.

J.W. Armstrong/Armstrong Photographic Trust

occurred very close indeed to the east portal of Deepgrove Tunnel, this was frowned upon. (See *Whitby Gazette* article, 5th December, 2014: *How Whitby escaped a major rail disaster.*) We have to remember, too, that these trains were full, packed to the limit, and the weight of 400 plus men as well as the steep gradients, sharp curves and slippery rails would mean that the motive power would have to be both tough and reliable.[1]

So a great opportunity to save the Whitby-Loftus line was lost, and it closed on 5th May, 1958. But what if it had not closed? With the opening of the new potash/polyhalite mine near Sneaton coming closer by the day, how easy – and cost effective – it would be to move the material by rail instead of building the unimaginably expensive underground pipeline, costing well over one and a half billion pounds, from the mine to Teesport. If the line were still open, material could be delivered to a railhead at Hawsker (as was planned in 1968 when plans for potash mining there were in full swing) and delivered by rail straight to the already functioning Boulby mine where potash is delivered every day to Teesport by rail.

Prospect Hill Junction in the 1950s; line to Scarborough (*left*).and to Whitby Town (*right*).
Author's collection

CHAPTER EIGHT

CLOSING A LINE BEFORE BEECHING: THE END OF THE WHITBY-LOFTUS LINE

Although it is arguable that the decline of the Whitby-Loftus railway began in *c.* 1925[1], it was not until the Transport Act of 1953 that the possibility of closure – remote at first – began to be more than a chimera. The previous Transport Act, that of 1947, while basically the instrument of nationalization in that it set up the British Transport Commission (BTC), while wishing to provide an 'efficient, adequate, economical and properly integrated system of public in land transport' nevertheless made little reference to costs, limited the railways' freedom of charging, continued out-dated statutory obligations and, more importantly for the Whitby-Loftus line, left the way open for cross-subsidization, where charges on profitable routes tended to be above marginal costs in order to subsidize the less economic parts of the network, and thus the retention of unremunerative services.'[2] The 1953 Act allowed the de-nationalization of long-distance haulage and limited the demands of the 1947 Act to merely 'providing a railway service for Great Britain'. Line closures had begun to increase: route mileage closed to passenger traffic was 343 miles in 1948-50, but by 1951-3 this had increased to 1,077 miles.[3]

Indeed, warning signs were apparent as early as 1952. It was in the October of that year that a census on the line was taken.[4] A letter of 3rd February, 1953 to the Chief Regional Officer and signed by the operating superintendent of BR (E and NE regions) at Marylebone, Mr E. W. Rostern, and the commercial superintendent at York (Mr E. W. Arkle) commented that 'when the special October census was taken the average throughout loading of these trains (the 11.40 Scarborough to Middlesbrough and the 1.7 Middlesbrough to Whitby Town) was 16 and 7 passengers respectively, with the former train reaching a maximum loading of 28 passengers on weekdays and 67 passengers on Saturdays approaching Middlesbrough'.[5] Unfortunately these figures do not indicate the loadings on the Loftus-Whitby section, although the average of seven on the train in the direction of Whitby (eight trains were used in the census) indicates a severe under-use. Even so, there was as yet no indication of any possibility of closure for on 5th March, 1953 the same two signatories (Rostern and Arkle) in a letter to the Chief Regional Officer (CRO) at York reached the conclusion that:

> In view of the heavy summer traffic it would seem that we must continue to provide a reasonable winter service and improve the loading by introducing

cheap fares. The possibility of re-arranging the service will be examined in connection with the Winter 1953-4 timetable.[6]

The line was, once again, being saved because of its exceptional usefulness during the brief summer season, and, given the emphasis of the letter on the introduction of cheap fares, the 're-arranging' may very well have meant additional trains. Because of the importance of the line in summer, especially in the number of passengers carried, it is worth considering in detail the operating changes made in the 1930s; it may also be argued that, firstly, because of the success of these changes, the line's survival to 1958 was assured and, secondly, that without these changes the line would surely have closed much earlier than it did.

The effects of changing the northern terminus of the line, in 1933, from Saltburn to Middlesbrough

Since the opening of the line in December 1883, the northern terminus of services had been Saltburn. Saltburn, apart from a few fishing cottages on the shore, did not exist before the 1860s; indeed, it is a railway town *par excellence*. However, the population has never been great. Thus the line, on its way to Scarborough, did not pass through any towns with a population greater than about 3,000 (except Whitby). It was decided – no doubt because of the immense fall in passenger numbers – to change to northern terminus of the line to Middlesbrough whose population in 1933 was *c.* 150,000. That this change was successful, in terms of passenger numbers, may be deduced, not from numbers themselves, which are not available after 1937, but by a document issued by the North Eastern area of the LNER.[7] The report began by stating that 'the figures in respect of passengers conveyed in the summer of 1933 had exceeded all records up to that time'.[8] The reader should be careful not to think that this statement referred to the Whitby-Loftus line; the traffic receipts for that year clearly contradict such a statement. It was the transference of the northern terminus to Middlesbrough which had caused such a growth. As such, the 'Coast Line' (at least in the summer season months) had been almost overwhelmed. The term 'Coast Line' had not appeared before and thus appears to be a marketing term. Again, the Whitby-Loftus section cannot be seen in isolation to the rest of the line to Middlesbrough and, indeed, it was (or so the report suggests) to enter into a period even busier than that of the early 1920s. This meant that the importance of the line was considerably increased;

its value to urban mobility, that is, that of the conurbations of Tees-side, Darlington and West Hartlepool (from which the summer timetabled trains began), or at least urban/holiday mobility was much enhanced.

However, success brought problems in its wake. It cannot be denied that the change to Middlesbrough as the northern terminus was a spectacular success. In 1933, in the summer season between June and September 273,098 journeys were made and during the same period in 1934, 294,855 journeys were made. Of these, in 1933, 168,848 tickets were collected at intermediate stations (almost certainly the majority being at Whitby (West Cliff), while in 1934 during the season 178,054 tickets were collected at intermediate stations (again, with Whitby (West Cliff) being the most likely destination). Total receipts from the line for these summer months were equally as spectacular. Including holiday season tickets, total receipts for the June-September 1933 season were £12,112, and for the same period in 1934, £16,475.[9] However, the operating improvement in 1934 was at some cost. In 1933 the line had been almost overwhelmed with traffic. Many trains ran late and there was a certain amount of chaos. It was the necessity to prevent further chaos in 1934, and, in doing so, to attract more people to the line, that the NE area report was made. That many improvements were authorized, at considerable expense, such as the construction of the new platform 1a at Scarborough (a dead-ended platform solely for use by Coast Line trains) which made it possible to deal with the increased service in a much more satisfactory manner than before, indicates that the LNER considered that the line was worth such heavy investment and, indeed, right up to the end in 1958 summer traffic remained heavy.[10] However, it must be emphasized that these improvements had very little effect, if any at all, upon the fortunes of the village stations on the Whitby-Loftus section of the line.

The improvements that the report suggested, and which were implemented, clearly were successful, in terms of the extra numbers of people attracted to the line and the improvements in punctuality. The report found that, in 1933, there was a serious problem with punctuality on the line. In July 1933, of the total of 356 trains on the line in that month, on average they ran 12.37 minutes late; in August the total of 331 trains ran on average 23.82 minutes late; in September 366 trains ran on average 7.68 minutes late. This meant that the main advantage that rail had over road, speed, was being eliminated. The report then made a number of suggestions to improve punctuality and the performance of the line generally. It was clear that the main problem was at Scarborough station, where the working of the Coast Line trains caused considerable problems to the efficient overall working there. The Coast Line trains had to make a reverse movement from the branch line into the station proper, thus often delaying other workings or causing even further delays to the service itself. The report proposed (and this was carried out before the 1934 season) that an additional terminal platform (1a) be made at Scarborough, close to the branch, and thus avoiding any crossing movements with other lines. The augmented service introduced in 1934 would have been impossible to operate without this improvement and, because it meant that the branch line was to all intents and purposes now self-contained, the improvement had a beneficial effect on station working generally.[11] This alteration was so successful that the punctuality figures improved dramatically. That the improvement was on a major scale is also indicated by the vast number of extra trains over the line in 1934. The punctuality figures for all trains terminating at Scarborough, Whitby, and Middlesbrough for July-September 1934 were as follows: July 1934, 608 trains were running on average 1.77 minutes late; August 1934, 571 trains running on average 3.48 minutes late, September 1934, 566 trains running on average 2.27 minutes late. Perhaps even more than the halcyon years of 1919-21, the years 1933-39 were the golden ones for the Coast Line in terms of numbers of passengers conveyed between the large Tees-side conurbation and the holiday resorts of Whitby and Scarborough. However, it must be emphasized again that this prosperity did not affect the stations on the Whitby-Loftus section, whose decline is shown by the traffic returns of those years.[12]

By 1934 the punctuality problems had been overcome, a vast augmentation of the timetable (in terms of the number of trains being run) had occurred, and the numbers of longer-distance passengers booking were very satisfactory indeed. This fits in with the wider picture nationally for these years, in that the longer distance trains were successful, while those slow trains which stopped at every station on a rural branch were losing passengers and income very quickly indeed. The United bus company put on huge numbers of buses on these holiday services, but it seems clear that the railway won this particular battle (once the 1933 upheavals had been sorted out) because of the advantages of speed and comfort. It is, though, unfortunate that it is not possible to make price comparisons. The rail timetables for the summer season of 1938 indicate how busy the line was, and it should be noted that these public timetables would not have included some relief trains. At the season's height there were 17 trains from Middlesbrough to Scarborough on weekdays and 11 trains on Sundays. In the opposite direction there were 18 trains on weekdays and 13 trains on Sundays. Leaving from Middlesbrough most trains departed between 6.30 am and 1.30 pm, 10 of which stopped at Hinderwell (a passing place) and three ran through non-stop. Returning from Scarborough, the majority of trains were in the evening, with 10 trains stopping at Hinderwell between 6 and 11 pm. In addition there were three non-stop trains.[13]

The Modernisation Plan of 1955 and its effects upon the line

However, even the success of the summer service (which continued after the war and, indeed, up to 1957, the last full year of operation) was not enough to halt the decline of the Whitby-Loftus section. On 24th April, 1954, a letter to the CRO at York considered the possible use of diesel multiple units. In view of decisions in 1956 concerning dieselization of lines to Whitby it is apparent that at this time experience with diesel working was limited and thus the thinking of this letter somewhat muddled. Nevertheless the future of the line still seemed secure:

> Lightweight diesel units could not cater for the summer traffic and different forms of traction for summer and winter must be uneconomic. The best prospects of more economic working lie in the direction of extending the Middlesbrough-Newcastle and possible Middlesbrough or Saltburn-Darlington schemes to take in these North Yorkshire branches, thus minimising costs occurred in overheads, spares, and maintenance charges. Except as an adjunct to another major scheme it would be difficult to find economic justification for putting diesel units on these (i.e. Whitby) branches.[14]

The beginning of the end for the line was marked by the publication of The Modernisation Plan of 1955. Increased investment and greater efficiency were to be at the cost of unremunerative services. Thus there was a growing realization of the need to accelerate the pruning of unprofitable branch lines. Indeed, the BTC stated that they were looking to undertake 'a substantial transfer of passenger traffic from stopping trains to properly designed road services'.[15] What was worse, from the point of view of those communities which seemed likely to lose their railway if these plans were carried out, were the guidelines given to the Transport Users' Consultative Committees(TUCCs). These guidelines, transmitted through the Area Boards in March 1956 had two main concerns: firstly, that arguments of objectors who claimed that branch lines could be kept open by alternative operating methods were to be countered; secondly, that the Minister of Transport was to be told that it was imperative that the TUCCs should not delay closure decisions by asking for additional information in cases where the financial justification for withdrawal of service was clear.[16] Both these elements were present in the discussions between the Whitby Rural District Council and the N.E. Area TUCC.

Things now began to move at a faster pace. A passenger census taken on 5th October, 1955 indicated that loadings were worse than in 1952. Again, eight trains were used in the census, showing an average number of passengers per train of 10. The 1955 census provided more details, estimating movement costs at £473, while giving receipts of £47. This showed an operating ratio of 1,006 per cent.[17] Coming so soon after the publication of the Modernisation Plan these were disastrous figures.

One of the main elements of the Modernisation Plan was that steam should be replaced by electric and diesel traction. The nationwide lack of investment and the enormous backlog of repairs, renewals, and maintenance were to affect the fortunes of the Whitby-Loftus line very seriously indeed. In April 1956, only a few days after the Area Boards' publication of ground rules to the TUCCs, British Railways announced its intention of converting to diesel traction all lines running into Whitby except that along the coast from Middlesbrough via Loftus. In fact, a letter dated 10th September, 1956 from the BTC (BR) (North Eastern Region) and signed by the chief commercial manager (Mr F. Grundy) and the chief operating superintendent (Mr A. P. Hunter) stated clearly what was to happen to the Whitby-Loftus line:

Name of Branch	Proposed action
Whitby-Loftus	Abandon line altogether when diesels introduced via Battersby.
M'bro-Whitby via Battersby	Multiple unit diesel trains to be operated.

✓ Middlesbrough – Whitby via Battersby		Multiple unit diesel trains to be operated. ✓
✓ Wakefield – Barnsley		Multiple unit diesel trains to be operated and remodel the service to give through trains Leeds – Wakefield – Barnsley. ✓
✓ Whitby – Loftus		Abandon line altogether when diesels introduced via Battersby.

Relevant part of the letter of 10th September, 1956, from the BTC.

BR (North Eastern Region)

This letter, which condemned the line, was certainly never made public and, indeed, the proposal for closure was not published for a further 12 months. Although not realized at the time, the April announcement was the death knell for the line. For despite remaining very successful in the few summer weeks of the season, the figures for out-of-season passenger traffic for 1956 were very sparse indeed, indicating that it would be very difficult indeed to justify the line's retention were it to be considered for closure.

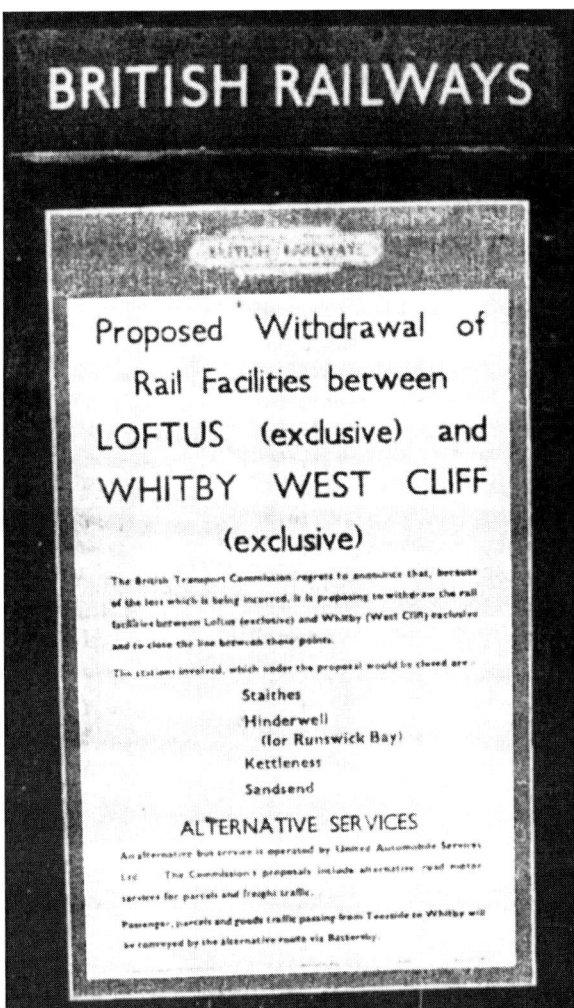

Official closure notice issued by British Railways.
J.W. Armstrong/Armstrong Photographic Trust

Closure

For the events during the last eight months of the line's existence, the most important and detailed primary source for the fortunes of the railway is that of *The Whitby Gazette*. Founded in 1854, the *Whitby Gazette* was, in 1957 and 1958, a weekly publication (Friday) which served all the communities through which the line passed. The standard of journalism and of the contributors was of a very high quality. In the second week of September 1957 the fatal proposal, perhaps not entirely unexpected, was reported in the *Gazette* in the September 17th issue. Amid strong protests of the Whitby Rural District Council (WDRC) it was stated that the closure of the line would be discussed by the North Eastern Area of the TUCC at Newcastle on October 3rd. The statements and protests of the WRDC were reported at some length. Alarm and dismay at the proposal were voiced with the Clerk of the Council, Mr J. B. McClurg leading and organising the opposition. But it was clear that the disquiet felt by the Council was tempered with resignation, for councillors recognized that the service was uneconomic. Councillor O. Welford said that sometimes he saw only five or six persons using the railway; Councillor G. Lyth was even more pessimistic, saying that he did not see any hope of the line remaining open, and even the Clerk admitted that 'one had to realize that the travelling public on the rail service between Loftus and West Cliff was not very great'. It was the isolation of Kettleness and Goldsborough which most concerned the Council, as well as the future transport of Whitby men who worked at the Skinningrove Iron and Steel works, some of whom still used the train.[18]

A week later the *Gazette* reported that opposition to the proposed closure was growing. The Whitby Chamber of Trade, the Whitby branch of the National Farmers' Union and Loftus Urban Council all entered the fray. One important piece of information which was given at the last Whitby RDC meeting was that the BTC (British Transport Commission) had found that it was not possible to use diesels (he meant the recently introduced diesel multiple units) on the Loftus-West Cliff line.[19] This, of course, begged the question which was shortly asked by Councillor J. A. Peace: if diesels could be run on some lines, then why could they not run on another. This question was unanswerable, the protesters never receiving any further information on this vital issue. There was, however, a growing sense of finality about the protests. Whitby RDC Councillor Owen commented that the £57,000 required for repair and maintenance of the viaducts and tunnels was not a large sum if the long-term benefits were taken into account.[20] However, it must be realized that this sum is the equivalent of almost £1 million in today's (2018) money.

One of the longest of the many *Gazette* reports on the proposed closure of the line was made in the issue of 11th October, 1957, eight days after the crucial meeting of the N. E. Area TUCC. The Clerk of the Whitby RDC, Mr McClurg had compiled a considerable portfolio of arguments to be presented to the committee. However, it is clear that, for all his good intentions, he was grasping at straws. There was no argument to be made against the need for expensive maintenance; the western portal of Deepgrove tunnel gave a clear illustration of what was required. Mr McClurg presented passenger statistics for the months January-August 1957 for Staithes and Hinderwell, but it is quite possible that these figures were counter-productive, producing the opposite effect from that which the Clerk might have hoped. For example it was shown that in March 1957 185 tickets were issued at Staithes. Trains did not run on Sundays, so the statistic would apply for 27 days, with six trains stopping at the station daily. This meant that seven passengers (to the nearest round figure) booked at the station daily, with an average of just over one passenger boarding each train. The figures for Hinderwell were worse. In the February of 1957 118 passengers were issued with tickets. There were 24 working days that month, again with six trains arriving daily at the station. This meant that five passengers booked each day, with an average of less than one passenger boarding each train.

Figures for Staithes January – August 1957

	Tickets issued	Tickets collected
January 1957	345	112
February 1957	304	135
March 1957	185	128
April 1957	218	275
May 1957	325	358
June 1957	2,057	2,271
July 1957	4,113	4,912
August 1957	3,067	3,073

Figures for Hinderwell January – August 1957

	Tickets issued	Tickets collected	Season tickets	Parcels received	Parcels forwarded
January 1957	155	236		178	15
February 1957	118	238		157	22
March 1957	142	257		154	23
April 1957	261	301		262	59
May 1957	203	478		234	45
June 1957	442	1,003	9	282	83
July 1957	1,551	2,110	16	261	106
August 1957	911	1,418	24	306	80

These figures were clearly disastrous and could not have advanced any argument put forward by Mr McClurg for the line's retention. He would have been advised not to adduce them. He was probably hoping that the summer figures, especially those of July, would come to his aid. Although 4,113 tickets were issued at Staithes and 1,551 at Hinderwell that July, this increase must be seen in the light of the increased number of summer trains stopping at the station. Trains ran on Sundays, so it would seem that, at Hinderwell, 50 tickets were issued daily, while at Staithes that number was a respectable 133 (rounded up). However, the great increase in the number of trains on the line during the summer season, 18 on each weekday, 26 on Saturday, and 12 on Sunday, meant that there was very little improvement in the number of passenger bookings at those two stations. Those were not convincing numbers and could not have aided Mr McClurg's opposition. Nevertheless 2,102 trains that month passed through the stations, some being non-stop.

Although Mr McClurg tried valiantly to prevent, or at least postpone closure, he was reduced to clutching at straws during this important meeting. Apart from the Staithes and Hinderwell figures he argued that the camping coaches at Sandsend/Eastrow, Kettleness, and Staithes brought in well over £5,000 p.a. His presentation to the N. E. Area TUCC was further weakened by a too-great emphasis on complaints about the service: too few trains, antiquated rolling-stock, and undue extravagance (he quoted an example of 16 men being sent from Darlington to repair a broken wire between Kettleness and Sandsend).[21] Thus it cannot have come as a great surprise when, on 18th November, 1957 the *Gazette* reported that the line was to close. Freight traffic was very light, passenger traffic was almost non-existent (apart from a few summer weeks) and, even if dieselization had been implemented, a large deficit would remain.[22]

From then, until the last week before closure, the issue became moribund apart from the occasional reporting of somewhat far-fetched ideas which the instigators hoped might spare the line. The MP for Cleveland, Mr A. Palmer suggested that the line could be run under the Light Railway Act[23], the Whitby RDC suggested keeping the line open as far as Hinderwell[24], but on 14th March the *Gazette* reported that the date of May 5th had been given as the closure date, with the last trains running on the previous Saturday, May 3rd.[25]

There appears to have been little or no protests from the general public who had long since deserted the line. There was silence on the pages of the *Gazette*, too, until the issue of 2nd May, 1958, which devoted 2½ columns to the incipient closure, almost entirely interviews with and stories about those who had, as children, taken the first train and on the morrow might well take the last. The central figure in these reminiscences was Mr William Pybus of Whitby, whose comments perfectly summed up the feelings of the travelling public over the past 30 years. He had, as a four-year-old boy, been taken on the first train from Sandsend to Whitby. However, when asked if he intended travelling on the last train he said 'I am not going to bother. Bus travel is far quicker. The buses are far handier, for you waste a lot of time getting to railway stations'. *Vox populi, vox Dei*. No statement could more adequately sum up the fortunes of the line since c. 1925.[26]

The last report concerning the line in the *Whitby Gazette* was made in the issue of 9th May, 1958. It is a masterpiece of journalism. The article, 3½ columns long and consisting of well over 2,000 words, deserves to be read in its entirety. The anonymous journalist was clearly moved by the occasion, recognising it to be an important moment in Whitby's history. Descriptions, interviews, reminiscences, and photographs commemorate the last trains on the line after almost 75 years of operation. One paragraph in particular, on the scene at Sandsend, sums up the sense of occasion and illuminates the quality of the writing:

> With the deep blue of the bay behind them, white capped waves beat over the nearby Promenade as people stood around admiring the view, and awaiting the arrival of the trains. Early holidaymakers in the camping coaches were also interested spectators. The sea thundered on the rocky foreshore below the station, gulls wheeled above the angry surf, smoke trailed behind two tiny coasters out on the horizon, and away to the left, towards Kettleness, the green banksides were starred with myriads of primroses. A beautiful scene from a vantage point never again to be enjoyed by rail passengers waiting for their trains in springtime.[27]

The penultimate train; the last train (5.28 pm) to Middlesbrough arriving at Sandsend.
Whitby Gazette

'Last scene of all, which ends this strange eventful history.'
The last train ever at Sandsend, the 5.44 pm to Scarborough. *Whitby Gazette*

THE END OF THE WHITBY-LOFTUS LINE

Staithes station in 1959 after the removal of the track.
J.W. Armstrong/Armstrong Photographic Trust

What is a railway for?

Thus the line passed into history. Weeds grew on the trackbed and on the deserted station platforms. The rails rusted, water dripped from the tunnel roofs, and birds began to nest on the columns and bracings of the viaducts. In 1959 the rails were removed.

Then, in May 1960 the unique and impressive viaducts were demolished. The service from Middlesbrough to Loftus limped on for a further couple of years, coming to an end in May 1960. West Cliff station closed to all traffic in June 1961, and from then on the line was forgotten. The tunnels were boarded up, road bridges were demolished, and farmers ploughed the line back into the earth from whence it had come.

On the face of it, the Whitby-Loftus line appears to be an almost complete failure, even from before its opening day. The investors lost money, the North Eastern Railway lost money, the Whitby, Redcar and Middlesbrough Union Railway Company was bankrupt and in complete disarray before the opening of the line, the original contractor and engineers were sacked, the designer of the viaducts was never fully paid, the line ran at a loss throughout its 75 year existence (excepting, perhaps, the years 1920 and 1921), freight was very limited – at times non-existent – and in the out-of-season months (perhaps nine) the timetabling was extremely sparse. We must therefore address an important question, one that is rarely acknowledged by railway historians: What is a railway for? Merely to provide profits for a company and its shareholders? Or to also provide a social service to its

Hinderwell station in 1959 after removal of the track.   *N. Cholmondeley collection*

locality and, if it connects with a main line, the country as a whole? Two influential historians approached this issue from opposing angles. R.J. Irving, who saw the entire history of the line as a 'spectacular failure' and T. Leunig, who considered that such lines not only increased 'social mobility' but also (unstated, but by implication) increased the well-being of the nation as a whole.[28] Irving's argument and my modification of it may be found in a recent edition of the *Journal of the Railway and Canal Historical Society*.[29]

Economic motives were, of course, of prime importance, but the building of a line was not necessarily entirely due to the desire for profit. Although the periods of railway 'mania' were now in the past when the Whitby-Loftus line came into being, by the 1870s the forming of new companies and the construction of new lines was still commonplace. New lines were, perhaps, *par excellence*, the visual evidence of a growing and dynamic economy. Towns which perhaps in the past had been resistant to the incursion of the railway now realized the benefits that a line could bring. The benefits of being the first industrial nation, which included expansion of international trade, the growth of the Empire, and the dominance of Britain politically in world affairs were a source of pride to the British middle class who formed the majority of investors to new railway companies. This pride was manifested locally in a desire to promote railways which might increase the wealth and status, not only of those who promoted and invested in a new railway company, but of the towns (and perhaps even villages) through which the line might shortly run. The expression of wealth and status, though an abstract concept in the sense that it cannot quantitively be measured, is one which should not be overlooked when considering the motives for the promotion and construction of a new line, especially one which might be considered rural, even remote. So who were these investors? A list of debenture holders in the Whitby, Redcar, and Middlesbrough Union Railway Company shows that the occupations of the shareholders indicate that they were a typical cross-section of provincial life (included were those who described themselves as farmer, builder, butcher, painter, clothier, and solicitor). However, the majority described themselves as 'gentlemen' (*see Appendix Six*).

Leunig's article deserves further consideration. It argued that although British railways benefited the economy and society in general, they benefited the passengers particularly. It reinforces the argument that the 19th century railways' social function was as important as their profitability.[30] It was the *time* saved by railway travel that was crucial: trains were much faster than coaches or walking; the cost of travel fell significantly, and its speed and comfort rose dramatically. As regards society in general, Leunig wrote that, 'The form of cost-benefit analysis used by historians to study railways is known as "social savings"'. Put simply, the social saving from railways is the minimum additional amount that society would have to pay to do what the railways did

Deepgrove tunnel after closure   *Author*

without them, that is, the cost of moving freight and passengers without trains'.[31] Economically, Leunig thought that 'the number of hours saved (that is, hours that could be spent at work) rose dramatically over the railway era, from 54 million hours in 1843, to 527 million hours in 1866, and finally to 5 billion hours by 1912, roughly equal to the hours worked by 1.8. million workers'.[32] Thus the railways (and here we must remember to include the Whitby-Loftus line) were part of a major transformation of British economy and society, so much so that, according to Leunig, 'The social savings from railways in time and money amounted to some 2 per cent of GDP as early as 1850, to 5 per cent of GDP by 1865, 10 per cent by the turn of the century, and fully 14 per cent by 1912'.[33] As for benefiting the passengers (both the middle and working classes, for third-class accommodation was used by many middle class passengers, certainly by the late 19th century), Leunig made the telling comment that 'people who could never have expected to travel in all their lives were able to do so for the first time. And those who did travel were able to do so more often'.[34] This was especially the case from the 1870s onwards, as railways sought to attract more customers with better third-class services (and, although Leunig does not say so, by implication the building of new lines). The railways may have offered poor returns to investors, concludes Leunig, but they delivered tremendous welfare gains to travellers and to society.[35] The Loftus-Whitby line, then, can just as easily be appreciated in a positive manner, in Leunig's terms, as in Irving's much more pessimistic view.

Even after the mid-1920s the line provided an essential service: that of transporting hundreds of thousands of city-dwellers from Middlesbrough, Hartlepool, Darlington, Sunderland, and even as far afield as Newcastle to the seaside delights of Whitby and Scarborough. Its initial purpose, though, of transporting ironstone, fish and agricultural produce to the industrial centres had long been abandoned, indeed if it had ever existed, while local bus services had effectively taken over the transportation of local passengers. But its main *raison d'être* after the mid-1920s was based on seasonal traffic – and that season was very short.

Nevertheless, after the final takeover by the North Eastern Company in 1889, it became part of the wider NER network and, mainly through cross-subsidization, was able not only to survive but to grow. As a project for an independent company and its shareholders, yes, it failed, but I believe that it played an important – or fairly important – rôle in the history of both railway and society in the north-east of England, its existence being based more on its social function than any other element.

CHAPTER NINE

THE RAILWAY IN THE IMAGINATION

The railway in the landscape. All lost and gone.
J.W. Armstrong/Armstrong Photographic Trust

Because something has vanished, because it can no longer be seen, doesn't mean that it's not there. So imagine the line on a May evening. But this is a May evening in 2018; dusk is falling, the last bus to Middlesbrough has just passed by as you are walking down Ellerby Lane, from Runswick Bay, to where the bridge over the road used to be. You can tell instantly where it is, not just by the dip in the road, not just by the familiar line of trees that often is the only visible clue to where a railway used to run, but by the indefinable romantic, mysterious, melancholy mood that always makes itself apparent when a forgotten line is nearby. There is a short path leading from the road up to a low embankment and, having negotiated tree roots and branches, you find yourself on a dry mud path almost overgrown with weeds, brambles, bushes, thorns, shrubs, trees, and all the flowers of the wilderness. It is a fairly well-worn path, indicating that others know of this hidden route to Kettleness. It is not a difficult walk, for the railway gradients, though sometimes steep to the train crews labouring their trains up a 1 in 60 incline, seem almost flat to the walker. It was here over the years that thousands and thousands and thousands of trains passed; here travelled millions of people on their way to the seaside, to work, to shop, to pray, to be born, to marry, to die; here travelled ladies and gentlemen in first class seats in 1884, here travelled soldiers and sailors on their way to war and death, here travelled successful authors on the way to their publishers in London, here travelled boys and girls on their way to school, some happy, some apprehensive, some desperate; here travelled youths on their way to the Empire cinema in Whitby on a Saturday night or a dance at The Spa on a Wednesday evening. All these ghosts passed over the earth on which you now walk; all alive, all human, all doomed. That is the poetry of history.

The trains passed in all weathers. You must imagine the line in winter. You must imagine the line in the landscape. It is a perfect, sunny, cold day in January. A train has just left Kettleness and is on its way to Whitby. The train consists of only two carriages. There are not many people aboard the train. Shortly after leaving the station the train passes into Kettleness tunnel. It has left a land of green fields, trees, hills, valleys, streams, villages, and farms. It is about to enter a very different country.

Everyone who ever rode on a train between Kettleness and Whitby will remember their astonishment as they left the short (308 yard) Kettleness tunnel. For instead of farmland the landscape has changed, magically and dramatically, to that of a wild, old, primitive country. To the left of the train the rock strewn ledge falls away sharply to the sea where, in January the sea roars and foams and flings itself with rage and fury against the cliffs. Today there is a ship, a coaster, wrecked upon the rocks below, there is a temporary waterfall fuming and falling hundreds of feet from the cliff top into the sea, the gulls are arching over the waves, and the wind is growing in intensity and strength. To the right of the train the cliffs rise to the sky, for the line is built on a shelf below these cliffs where it runs for perhaps 300 yards before disappearing into the mile long Deepgrove tunnel. The near gale rattles the windows of the antiquated carriages and the steam from the sturdy 'A8' tanker vanishes into infinity. The train has gone. The line remains. The wild coastline retains its tremendous power. This is the railway in the landscape. This is the spirit of England.

CHAPTER TEN

WHITBY'S THIRD STATION

Almost outside the Community College, just before the Mayfield Road traffic lights which direct traffic either to Whitby, Scarborough or Teesside there are two tall barriers on either side of the road which prevent anyone looking over them. If they could see above these barriers, people would be able to look at what was once the most dramatic of all 19th century earthworks constructed in Whitby. Today, these earthworks have been overgrown and have become part of what is known as 'The Cinder Path' (*below*), a 20 mile cycle and walkway between Whitby and the edge of Scarborough. Sixty years ago, though, someone peering over the barriers would have seen a completely different view (*facing page*).

In May 1881 only a few miles of the Whitby-Loftus line had been completed, and even less of the line between Whitby and Scarborough. Both lines were run by independent companies at that time and they were very aware of each other's progress. Although separate companies, it was suggested that a combined station be built at Whitby. But where was it to be? As yet, West Cliff station had not been constructed and

Prospect Hill Junction today. *Author*

The likely site for Whitby's third station. Mayfield Road bridge is in the background, with the grounds of Airy Hill on the right. *J.W. Armstrong/Armstrong Photographic Trust*

although the 1881 Prospectus for the Whitby-Loftus line described its possible future construction as 'a station in a populous suburb of Whitby' this was not quite true, for the area was not a 'populous suburb' in 1881 and, indeed, maps published as late as 1925 showed the locality as remarkably undeveloped.

It was not long after the publication of this Prospectus that the Directors of the Scarborough and Whitby railway had received a request from 'the ratepayers of Whitby' that a railway station be built at Larpool. It seems reasonable to suggest that this proposed station would have been on the south bank of the Esk above the road that leads from Whitby to Ruswarp, adjacent to the viaduct which was at that time as yet unbuilt. The Scarborough and Whitby Directors refused this request on the grounds that the location for this proposed station was situated on too steep a railway gradient. As the gradient at this location was 1 in 43 (extremely steep for a railway) the company's refusal was not unreasonable. Then, in 1882, the Scarborough and Whitby company instructed their land agent to buy 'sufficient land at Whitby for a small independent station'. No doubt this would have expanded into the combined station which the two companies wanted. As the proposed station at Larpool would have served only the Scarborough and Whitby railway, and as the proposed station at West Cliff would have served only the Whitby-Loftus line, it seems obvious that a station adjacent to Prospect Hill Junction could be the only location possible. Thus it was here that Whitby's third station would have been built.

Ease of access would not have been especially difficult and two platforms could easily have been built. If this particular location was found difficult to construct, then there was ample space to the west (the other side) of the Mayfield Road bridge. While both the proposed West Cliff and Larpool stations would have been some way from built-up areas, this site at Prospect Hill was convenient for the town.

However, the next chapter in the story put paid to any likelihood of a third station for, in 1881, the contractor in charge of the construction of the Whitby-Loftus line (John 'Paddy' Waddell) was awarded a contract to build six stations on the line, the architect being William Bell. West Cliff station was built in 1882 and the proposed station at Prospect Hill forgotten. The Whitby-Loftus line opened in 1883 and 18 months later the line from Prospect Hill to Scarborough opened. Through running from Saltburn to Scarborough was instituted. This meant that anyone wishing to travel to the main Whitby station in the centre of town – then called Whitby (Town) - would have to change at West Cliff and either undertake a longish walk into town or await a shuttle service. So, had the station at Prospect Hill been built there would have been no need to change trains and for the inconvenient shuttle service. Passengers would have had an easy walk into town.

So overgrown has the locality become it is difficult to realize that Prospect Hill was once the centre of Whitby railway activity. In the summer trains would pass through the junction with great regularity and from very early in the morning until late in the evening. The sound of a steam engine blasting its way up from Bog Hall or whistling as it approached the signal box were familiar sounds. As holidaymakers and locals wait today for overcrowded buses to Scarborough or Middlesbrough one realizes the folly of closing these remarkable lines which ran along the coast.

Another view of the likely site for Whitby's third station. Prospect Hill Junction looking towards Mayfield Road bridge (in the background). *Author's collection*

View today from underneath Mayfield Road bridge toward Prospect Hill Junction where the likely site for Whitby's third station was situated. *Author*

CHAPTER ELEVEN

THE STATION THAT NEVER WAS

The two Prospectuses issued by the Whitby, Redcar, and Middlesbrough Union Railway in 1871 and 1881 both spoke in glowing terms of the benefits of tourism.[1] Apart from Sandsend and Whitby the greatest attraction for tourists along the line, then and now, was Runswick Bay and so, for 75 years, tourists alighted at Hinderwell. According to *The Official Guide to Whitby* of 1934, travellers, once they had alighted 'must take the footpath across the field immediately after leaving the station. This leads into the road which leads directly down to Runswick'.[2] The distance is approximately a mile and a quarter from station to beach. Of course, this kind of distance was no obstacle for tourists, although those with young children might very well find such a journey tiring. It was proposed, then, that a station should be built at Runswick to facilitate the trip to the beach.

Historians are like detectives. Railway historians perhaps even more so. The unexplained must be explained, hidden secrets must be discovered, gaps must be filled in and mysteries must be solved. The railway historian, researching the now elusive Whitby, Redcar and Middlesbrough Union Railway (now known as the Whitby-Loftus line) will come across hidden viaducts, long-lost tunnels, near-miss disasters, forgotten signal boxes, impossible plans (like the one to construct the line around the cliff edge between Sandsend and Kettleness) and memoranda which, for example, explain why the line took so long to build and was so costly. However, no researcher has been able to write about the station that never was until now. This is because the key document concerning this station had been inserted in a folder along with other, larger, plans and had been completely overlooked. As far as I am aware only two railway historians, Ken Hoole and myself have undertaken detailed research into the line and it may be that Hoole, although clearly having visited the National Archives at Kew, did not venture into the upstairs large document repository in that building. While researching the line I did on three or four occasions visit the large document repository and, during one of those visits, I came across – quite by chance – this key document, which seemed to have become attached to the underside of a larger document and, had it not been disturbed by my examination of the larger document, would have remained unknown. In a sense this serendipitous occurrence lies more in the realm of metal detectorism than historical research, for its finding was by chance, not by intent.

The document is a station plan, exactly like all the other station plans for the proposed Whitby, Redcar, and Middlesbrough Union Railway, but it is different in one vital respect: while all the other stations were built, this was not. The document is the proposed plan for Runswick station, which was to be built on both sides (east and west) of the bridge crossing Ellerby Road. Only on the south side of the line was there to be any construction; the plan was for one platform only. The site of the proposed station can clearly be seen on the O. S. map on page 21, on the minor road that runs south-west from Runswick.

The document measures 2ft 11in. by 1ft 2½in. It is in fairly good condition, though somewhat faded and yellowed by time, also there are four small holes which do not interfere with any reading of the item. The National Archives reference number is RAIL 743/20. The document is exceptionally detailed concerning information regarding the line and the proposed station. The line approaches from Kettleness on a gradient of 1 in 60. The platform was to have a 15 feet ramp at both ends and was to be 31 inches above rail level. Its length was to be 250 feet and, approximately halfway along a small booking office was to be provided. Fieldwork indicates that the platform was approximately 12 ft above road level and, although it is not shown, a path or road would have had to be constructed from the eastern side of Ellerby lane to connect with the station. The platform was entirely situated to the east of the road bridge. The gradient lessened to 1 in 200 as the line passed through the Hinderwell end of the platform, but immediately returned to 1 in 60 after the road bridge was passed. Within a few yards of the bridge, at the west (or Hinderwell) side a short loop was to be constructed, presumably to stable carriages on excursion trips to and from Runswick Bay. Immediately to the south of this loop there was to be built an approach road from Ellerby lane (with a 1 in 50 gradient). This seems a little odd, as the platform was on the other side of the road bridge (where an approach road would have had to be built, but there is no indication of this on the plan). The approach road evident on the document must have been planned as ingress for carriage cleaners or for locomotive maintenance. There is no evidence whatsoever of any facilities for passengers by the loop.

The document is dated 29th December, 1873 (quite late in the history of the construction of the line); as this is immediately prior to the dismissal of the original contractor (John Dickson) for the WRMUR and a few months before negotiations began for the takeover of the line by the NER, it may be assumed that Runswick station was a casualty of the demands which were made to the WRMUR by the NER as a pre-requisite for taking over the cost of construction the line. Finally, the document states that the platform was to be 2ft 3in from the inside of the rail. A station here would certainly have cut down walking time to the village at the bottom of the hill, which would have been – and still is – a popular destination. On the 1955 O. S. map it is still evident that those wishing to walk to Runswick from Hinderwell station would have had to have taken the footpath across a field (I remember this well) and then undergone the fairly long walk to the sea. Returning up the very steep hill to Runswick bank top and then have to walk back to Hinderwell station must have been a daunting task for the elderly or those with children. Nevertheless it is obvious why the station was never built: it is just too close to the much larger station at Hinderwell, and, furthermore, would only have been used to its full capacity for at best three months of the year.

The document

TNA RAIL 743/20 is not particularly easy to reproduce. The image below shows a very reduced document, the sketch shows the important details of the proposed station:

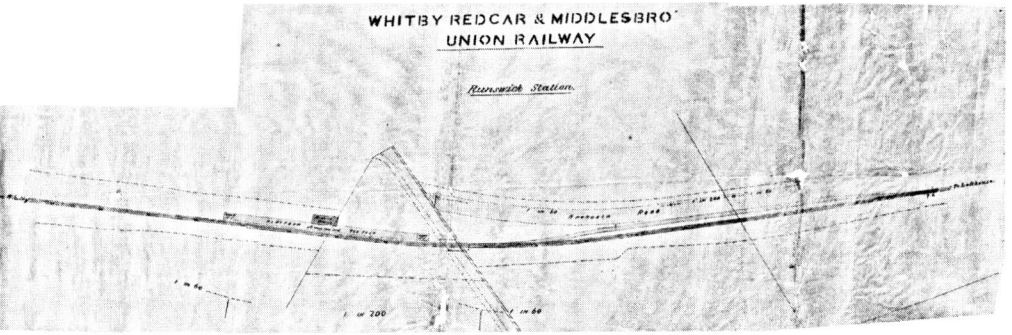

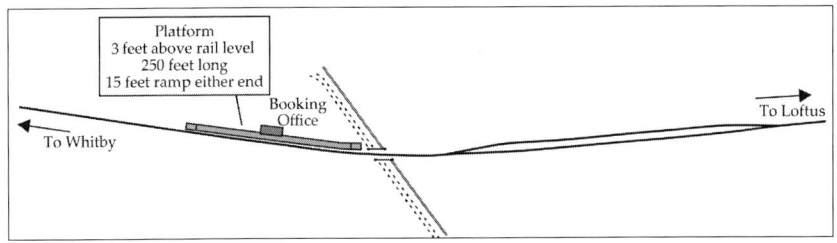

APPENDIX ONE

Transcript of the T. E. Harrison Memorandum of 14th Nov., 1883

The North Eastern Company became lessees of the Whitby, Redcar, and Loftus Railway 1st July 1875, but by the Act of 1876 for a deviation of the line, the time for the completion of the works was fixed for the 13th July 1881. The Company agreed to complete the line at the North End from Loftus to Easington 1½ miles, within a limited period, and these works were commenced in November 1876 and completed July 1878 under the contract let to Mr. Ridley. No time was lost after the passing of the Act of 1876 for deviating the line in preparing the plans and specifications for letting the works for the completion of the remainder of the line from Easington to Whitby.

All the bridges with the exception of one were so defective, and in such a dangerous state, that three were obliged to be taken down and rebuilt before the contract was let, and plans had to be prepared for rebuilding or strengthening the others, and nearly every abutment of the viaducts had to be taken down and rebuilt. The culverts were in many cases inadequate to carry off the water. The slopes of the cuttings and embankments were in no case sufficient, the consequence being that slips had taken place to a great extent, and the big cutting at Whitby (*presumably Prospect Hill*) was filled with slurry to the depth of 12 feet. This state of things coupled with the fact that there were no plans or sections that could be relied upon necessitated a complete re-survey of the whole line, and cross sections of every cutting and embankment, so far as they had been executed, a labour far greater than would have been required had no works been executed at all. It may here be stated that in some cases the surveys were so inaccurate that had the original tunnels been completed, they were so out of line with each other, that no junction could have been made with them at all.

T. E. Harrison, Esq., Chief Engineer, North Eastern Railway Company.

The contract, however, was let to Mr. Waddell in November 1878 and the time fixed for completion was 1st May 1881. The contractor was unable to complete the works by May 1881 owing to two very severe winters, which rendered it impossible to carry on the works in the half finished state of the cuttings and embankments which were full of slurry, and in many cases the work was obliged to be laid in entirely during the winter. There also occurred during the execution of the works, some of the heaviest slips I have seen, one at the south end of the Easington tunnel requiring a very heavy retaining wall and the removal of 35,000 yards of excavation, and a land slip a little further south, extending over 4 acres of land which slid away for 300 yards. During the excavation of the ordinary work of cuttings, embankments and bridge building, it was impossible to do anything to the viaducts, as they were constantly in use for carrying materials etc. over them, and in the case of the Staithes viaduct, nothing could be done to it until the abutments were built, and the embankments at both ends made up, and then it was required for carrying over it a large quantity of excavation from near the Easington tunnel and ballast for the railway.

The roadway for the permanent way over the viaducts was of the flimsiest character and such as no Govt. Inspector would under any circumstances have passed, and an entire new roadway was provided for all the viaducts as soon as it could be done without stopping the other work. It was considered necessary also to provide for the Staithes and Upgang viaducts two rows of longitudinal bracing to prevent buckling of the high piers, and this was done without delay. In July 1881 however a new set of requirements was issued by the Board of Trade which specially affected the viaducts, and for the first time it was required: 'that the viaducts should be equal to withstanding a wind pressure of 56 lbs to the foot'. The viaducts as built were only calculated to withstand a wind pressure of 28 lbs to the foot. This required a great additional strength to be given to the viaducts, and as they had not been constructed with such a requirement in view, it was a difficult matter to design the requisite alterations, and it was only after much thought that a design was prepared and carried out which has since been passed by the Govt. Inspector.

John Dickson, Esq., the original contractor who was sacked by the Whitby, Redcar and Middlesbrough Union Railway Company in December 1873.

A second new regulation was made to the following effect: 'If in iron viaducts the main girders are placed below the level of the rails substantial parapets about 4'6" in height must be provided, and as a further protection substantial guards should be fixed outside, above the level and as close to the rails as possible….' In making the new roadway over the viaducts check rails had been provided such as would have fulfilled the old requirements. It was thought desirable to ascertain from the Govt. Inspector who would eventually have to pass the works, what would satisfy these requirements, and after an interview with the Inspector in May 1882 in London and an application to the Board of trade an unofficial inspection was made and the requirements settled and they were immediately proceeded with.

These requirements however were of a very heavy character, and took considerable time to carry out and at considerable cost. In October 1882 the usual notice for inspecting the works with a view to opening the railway was given, and the works were inspected by the Govt. Inspector, his inspection on that occasion confined to the viaducts, for finding he could not pass them he deferred further inspection. The Inspector required that in the iron casing of the piers of the Staithes viaduct, holes should be cut in them to ascertain the condition of the concrete with which they were filled, the result being that in the first trial place the concrete was found to be mere gravel without any cement, and with the same result, though not so bad, in several other cases. It was further ascertained that several of the piers were not perpendicular, in one case to the extent of 7", in others to the extent of 3" or 4", and this applied more or less to all the viaducts. When this state of the work was discovered it was not thought desirable that any formal report should be made by the Govt. Inspector and the 10 days notice was accordingly withdrawn. It was then arranged that steps should be taken for a complete examination of every pier in each viaduct with reference to the concrete, and each pier that was out of perpendicular was with great difficulty straightened.

It has been asked how it occurred that the Engineer-in-Chief had not discovered these defects before. The answer is simply that he did not believe that such scamping of work could take place with even reasonable inspection, and such a case had never come to his knowledge before. No time was lost in remedying these defects and by aid of a force pump machine designed for the purpose, liquid cement has been forced into every pier, 1600 bushels of cement having been used, the result being that the viaducts have at last been passed by the Govt. Inspector.

It would be difficult to find anywhere a work so thoroughly scamped from end to end as this was when taken possession of by this company. Viaducts and ordinary bridges badly designed and as badly executed, and in a dangerous state. The driftways for the execution of the Easington tunnel had been so slightly stayed that they had all fallen in, and this made the construction of the tunnel much more difficult, all the cuttings and embankments too steep slopes and no drainage in any case. Permanent way not such as would have been passed, and the line could hardly have cost more if nothing had been done to it. The amount spent in putting the viaducts alone into a proper state exceeded £30,000, but for the work to the viaducts the line would have been completed nearly 18 months earlier.

Thos. E. Harrison
Newcastle
November 14th, 1883.

APPENDIX TWO

Station Receipts 1897-1907

APPENDIX THREE

DEEPGROVE TUNNEL DERAILMENT
LNER (NE area)
REPORT OF ACCIDENT

Kettleness Station   2nd. August, 1927

Date: 31/7/1927   Time: 9.33 p.m.   Place: 200 yards E of Sandsend Tunnel mouth

1. Nature of accident and upon what line: Derailment. Single line.
2. No. and desc. of train concerned; where from and to: No. 70. West Hartlepool-Whitby return excursion.
3. No. and class of engine: Engine 1459 (J 39)
4. Names of driver, fireman, and guard: Driver J. Harkers (W. Hartlepool), Fireman Hildrick (Whitby), Guard G. Spensley.
5. Lines blocked, how long, what trains detained: Main line from 9.33 p.m. 31st July 1927 to 4.15 p.m. 1st August 1927
6. If single line arrangements were made, give particulars and who supervised the working: Yes! Up to obstruction on both sides. I supervised the working on this side.
7. Particulars of all damage done. Nos. of damaged vehicles and how ticketed: Engines J39 1459 and O 1914. Coaches 2187, 13647, 3057Y, 21720, 2527, 2246.
8. Cause of accident – any blame?: Unknown

APPENDIX FOUR

REPORT OF ACCIDENT

About 10 p.m. my signalman informed me there had been an accident in the section and on coming out I ascertained from Sandsend that the guard of the return excursion from Whitby to West Hartlepool had returned there and that the train engine and some coaches were derailed about 100 yards from Sandsend Tunnel mouth. The guard stated two women were complaining of injuries and I arranged that the Sandsend S.M. should see to this and report all circumstances to you.

I wired at 10.2 p.m. – after confirming with Whitby Loco. Foreman – for the Middlesbrough tool vans to attend and I wired particulars to you and D.E. Darlington at 10.30 p.m.

It was decided to get an empty train from Saltburn and transfer the passengers on the spot. Saltburn was wired and at 11.30 p.m. an empty train consisting of 8 bogies left Saltburn.

I arranged with SM Sandsend to call his platelayers to protect the obstruction and lay detonators and also to send word to the fireman to bring the tablet to the station immediately. This was done.

There was only the Ganger at home here so I called him out and appointed him pilotman from this station to the point of obstruction.

The empty train from Saltburn arrived here at 12.53 a.m. and was split up into two portions of 4 coaches each. I took the first portion down to the disabled train at 1.13 a.m. and on arrival I found that assisting engine 1914 had one leading wheel off and the train engine – 1459 – had one leading wheel and two tender wheels off [*excursion was double headed [J39 and G5], my italics*] and that 6 coaches all had wheels off also. The remainder of the train – two saloons and a brake third – had remained on the rails. [*i.e. a 9 carriage train – my italics*]. Fortunately, all the derailed vehicles kept 'head on' none 'slewed' across the track. The derailment appeared to have taken place 200 yards east of (before reaching) the tunnel mouth and to have travelled some 120 yards to a point 80 yards from the tunnel entrance and the permanent way for this portion was very damaged and 'spread' – all chairs etc. at one side being broken.

After transferring about half the passengers from the excursion to the empty train and seeing that the obstruction was being properly protected and that the man in charge of the obstruction had signed the pilotman form I returned to Kettleness with the loaded four coaches at 1.50 a.m. and then took the other four coaches back for the remainder of the passengers. On arriving at Kettleness the train was joined up and despatched on its journey to West Hartlepool at 2.50 a.m.

At 3.13 a.m. the tool vans arrived and after I had described the state of affairs they left the crane here and proceeded to the scene of operations. On arrival it was found that the Whitby tool vans had reached the obstruction and had got engine 1914 righted and the Middlesbrough and Whitby men there set about getting the 2nd engine on the rails. When the two engines were re-railed they were towed back to Kettleness arriving at 4.47 a.m. and work was proceeding on the derailed coaches.

About this time the engine which had entered the section from the Whitby end took back to Whitby the 2 saloons and brake that had stayed on the rails.

Two of the carriages that had been derailed were – when re-railed – taken back to Whitby and the remaining four along with the tool vans arrived here at 12.22 p.m. on the 1st. Aug.

Then at 12.43 p.m. a permanent way [spel?] with the material necessary to repair the damage left for the scene of the accident, and at 4.15 p.m. the Per. Way Dept. advised me that trains may be allowed to pass through at caution.

The two pilotmen then came through to this end of the section with their forms which

I cancelled and I allowed the 3.12 p.m. from Saltburn to go forward to West Cliff at caution and ordinary working was resumed.

With regard to the passenger train working arrangements the Whitby S.M. informed me you instructed him to arrange for motor buses to transport passengers for Kettleness to Whitby and vice versa. This arrangement worked well up to dinner time but in the afternoon when the volume of passengers was highest unfortunately the supply of buses was lowest and it caused considerable friction.

The 7.9 a.m. from Saltburn was run as far as here and on arrival here I sent it back to Saltburn to take the place of our 7.44 a.m. leaving here 8.31 a.m.

The next train to arrive was No. 52 – the 10.22 a.m. relief ex Saltburn. This was sent back immediately to Saltburn with instructions to stop at every station leaving at 11.47 a.m. At 12.8 the ordinary 10.35 ex Saltburn arrived and after dealing with passengers and luggage I wired S.M. Hinderwell that I was sending this train away empty to him and that he had to hold it there to work forward in the times of No. 22 Up. The empty train left here at 12.19 p.m. This train was sent to Hinderwell to stand because I was crowded out here with damaged stock – engines, tool vans, and per. way wagons etc. I informed West Cliff and Whitby in case they had passengers to send them to Hinderwell.

The next train to arrive was the 1.0 p.m. from Saltburn which arrived here at 2.16 p.m. I held this train here to work back to Saltburn in the times of our 2.58 p.m. It got a late start – leaving at 3.15 p.m. – because I had word from Whitby and West Cliff that 20 odd passengers were coming out by bus and so I waited for them.

No more passenger trains arrived until the 3.12 from Saltburn which went through to West Cliff.

I enclose herewith extract from Block Register together with cancelled pilotman working forms.

(signed) W.R. Readman

APPENDIX FIVE

Whitby Gazette, 5th. August, 1927.

EXCURSION TRAIN OFF THE LINE

The first portion of an excursion train returning from Whitby by the coast line to Hartlepool, at about half past nine on Sunday night, left the metals on the Sandsend side of the long Kettleness tunnel (*the writer means Deepgrove tunnel*). The train was a heavy one, and two engines and six coaches left the line and damaged the permanent way for a considerable distance. As it was impossible to remedy the accident within a short time, a relief train was despatched from Saltburn, and the stranded passengers were despatched on their journey about two o'clock on Monday morning. A few people who had preferred to walk back to Sandsend spent the night there. A large force of platelayers was engaged in the repair of the permanent way and it was possible to resume the normal passenger service at four o'clock on Monday afternoon. Mr. W. B. Sinclair, stationmaster at Whitby, and his staff made every effort to deal with the difficult position, and with considerable promptitude steps were taken to deal with the heavy holiday traffic on Monday morning. Arrangements were made with the Pioneer Bus Company and the Whitby Motor Transport Company for a fleet of motor-buses and coaches to run between Whitby and Kettleness, calling at West Cliff and Sandsend and by this means a through service with Saltburn and the North was maintained until normal railway facilities were resumed.

APPENDIX SIX

OCCUPATIONS OF WRMUR
DEBENTURE HOLDERS
(TNA RAIL 743/5)

List of Debenture holders falling due during the half year ending 30th June, 1878

Banker	2
Merchant	1
Esquire	2
Soap manufacturer	1
Gentleman	3
Solicitor	1
Brewer/Maltster	1
Wool merchant	1
Clerk (i.e. clergyman)	1
Widow	3

As at 31st December, 1878

Surgeon	1
Spinster	2
Wife	2
Gentleman	8
Widow	2
Not given	3

List of Debentures falling due
1st January, 1879 – 31st December, 1880

M.P.	1
Bankers' clerk	2
Lt. Col.	1
Gentleman	13
Builder	1
Farmer	4
Plumber	1
Spinster	8
Surgeon	2
Professor of Music	1
Widow	3
Esquire	4
Wife	2
Clerk (i.e. clergyman)	3
Butcher	1
Actuary	1
Not given	4
Merchant	1
Solicitor	1
Corn Miller	1
Seed crusher	1
Clothier	1
Book Keeper	1
Painter	1

APPENDIX SEVEN

RAILWAY INSPECTORATE: INSPECTORS' REPORTS (1883).
(TNA MT29/44)

Transcripts of these three reports.

Report dated 7th July, 1883. From Major-General C. J. Hutchinson, RE.

Sir,
I have the honour to report, etc....

The line, which is a single line, 16 miles 27 chains in length, is an extension of the North Eastern company's line, terminating at Loftus, from Loftus to Whitby, at which latter place it joins the line between York and Whitby, a short distance from Whitby station.

It was commenced many years since by an independent company, and has now, with certain authorised deviations from the original plans, been completed by the North Eastern Railway company, into whose hands it has now practically fallen.

The permanent way is of the type now employed on all the North Eastern Railway company's new lines.

The steepest gradient has in inclination of 1 in 50 and the sharpest curves (of which there are several all provided with check rails) a radius of 10 chains.

The new stations on the line are Easington, Staithes, Hinderwell, Kettleness, Sandsend, and West Cliff (the station for Whitby). These stations have all been provided with the necessary accommodation, and the signal arrangements have been (with certain exceptions) properly carried out. At the junction with the York and Whitby a new signal cabin has been erected.

The works on the line are very heavy and consist:

1. 18 bridges over the line; all constructed with masonry abutments, 10 having arched tops (widest span – 40 feet, 6 wrot [sic] iron girder tops (widest span – 41 feet) and 2 timber tops (widest span – 25 feet).
2. 15 bridges under the line, all having masonry abutments, 7 have arched tops (widest span – 12 feet), 5 have wrot [sic] iron girder tops (widest span – 30 feet) and 5 have cast iron girder tops (widest span – 25 feet).
3. The viaducts, all having stone abutments, with intermediate piers consisting of cylinders of wrot [sic] iron boiler plate, varying in diameter from $4\frac{1}{2}$ feet to $2\frac{1}{2}$ feet and in height from 152 feet to 39 feet, filled with cement concrete, on the top of which, and not on the boiler plates, the superstructure is intended to be carried; the first [Staithes] of these viaducts (5) has 6 spans of 60 feet, in which warren [?] girders, and 11 spans of 30 feet, in which plate girders have been used; the second [Sandsend] has 1 span of 36 feet, 5 of 30 feet, 1 of 28 feet, and 1 of $27\frac{1}{2}$ feet, in all of which plate girders were used; the third [East Row] has 5 spans of 60 feet, 1 of $57\frac{1}{2}$ feet, in which lattice girders, and 2 of 28? feet, in which plate girders have been used; the fourth [Newholm] has 9 spans of 30 feet, and 2 spans of $27\frac{1}{2}$ feet, in which plate girders have been used; and the fifth [Upgang], has 5 spans of 60 feet, and 1 span of $27\frac{1}{2}$ feet, in which lattice girders have been used.
4. 3 tunnels, all lined throughout either in stone or brickwork, 990 yards, 308 yards, and 1649 yards in length.
5. 8 large masonry culverts, widest 8 feet.
6. Several retaining walls at the sides of cuttings, and underpinning cliffs.

These works (with the exception of the viaducts) appear to have been substantially constructed and, with the exception of a settlement in the abutments of the bridges at 13 miles 35, and 15 miles 21 chains, which should be carefully watched, to be standing [....]the girders in all cases have sufficient theoretical strength, and gave moderate deflections under test.

The construction of the piers of the viaducts is of a very peculiar character as left by the original company; they (especially the high ones) were very insufficiently braced, and much has since been done to improve their stability. I have made three several inspections (the first last November) of them, and various suggestions which I have made from time to time have been duly carried out, the condition of the concrete is thus the remaining problem which is now a subject of various considerations between the Engineer and myself, and he is about to remedy certain defects which showed themselves at the last inspection.

As regards vibration and oscillation, the viaducts now behave well with heavy engines passing over them at speed.

There are two authorised level crossings of public road provided with proper gates etc. and there is also at Bog Hall junction, a crossing of what is stated to be a public road, and which, if so, is unauthorised.

The fencing is of post and rail, and post and wire.

There are several deep cuttings and high embankments, where much trouble has been experienced from slips.

The following are the requirements:

1. Staithes viaduct – the longitudinal bracing should be extended from 3 spans, the ranging of the girders should be as far as possible improved
2. In all the viaducts the condition of the concrete requires careful examination, and means taken to improve it where defective
3. Means should be taken to prevent the undermining of the sea cliff on the north side of Sandsend station
4. At Loftus station the normal position of **B** points should be for the runaway siding. The dead ends of the runaway sidings should be provided with interlocked safety points
5. Hinderwell station: No. 1 lever should be interlocked with **B** and **C** points, and No. 5 with **A** points. A locking bar requires lengthening
6. Kettleness station: The down starting signal should be moved to the up side of the crossing; the normal position of point (9?) should be reversed. Safety points are required on the siding joining the down line. A locking bar requires lengthening; the safety points of the siding joining the up line requires turning away from the main line
7. At Sandsend station the parapet on the viaduct should be heightened on each side for about 100' in length. No 5 signal should be interlocked with **A** and **B** points. No 6 signal should be dispensed with and No 2 signal (altered in position) should apply to the depot. For the present the use of the loop line should be dispensed with. The normal position of the facing points should be for the depot.
8. At West Cliff station: a footbridge or subway should be provided. The normal position of points **A** and **D** should be reversed; safety points should be provided on the runaway siding joining the down line. A locking bar should be lengthened.
9. Staithes station: The siding at the viaduct end of the station should be taken out.

Numbers 4 and 6 should be interlocked, and points **D** should be interlocked with signals Numbers 1 and 4. The parapet on the viaduct should be lengthened for about 100' on each side.
10. Easington station: The safety points should be provided with wheel blocks. No 1 signal should be interlocked with (5?) and 4. 4 should interlock with **A** and **B**.
11. Bog Hall junction: No 2 should lock 17 in position, the distant signal repeaters should be placed opposite the levers. If the level crossing is that of a public road, the junction points must be removed from the crossing and proper gates, worked from the cabin, be provided.
13. Flange ties are required at several facing points.

....I cannot recommend that the opening of the line should be sanctioned and I must report that by reason of the incompleteness of the works, it cannot be opened without danger to the public using it.

C. J. Hutchinson, Maj. Gen. RE.

Report of Major Marindin. York, August 22nd, 1883.

Sir,
I have the honour to report for the information of the Board of Trade that in compliance with the instructions contained in your minute of the 11th instant, I have inspected the WRMUR branch of the NER, and with reference to the requirements noted by Maj. Gen. Hutchinson in his report of 7th July, 1883.

I find that the whole of the minor requirements have been complied with and that a footbridge for Cliff end [sic] station has been ordered and I was informed with reference to requirement #11 that the crossing at Bog Hall junction has been ascertained to be that of an occupation road only.

The following requirements have not yet been satisfied, viz:

#1 the ranging of the girders in Staithes viaduct cannot well be altered and Maj. Gen. Hutchinson has informed me that he does not attach great importance to this being done.
#3 nothing has been done as yet to improve the underpinning of the cliff at the north end of Sandsend station and the Engineer, although quite prepared to do whatever may be considered necessary was under the impression that no immediate action at this point was required by Maj. Gen. Hutchinson.

It might in my opinion be deferred for the present – as there is no sign of any danger to the line.

With regard to requirement #2, the concrete filling of most of the piers has been found to be very defective and four of the columns in each of the five viaducts were selected by Maj. Gen. Hutchinson to be dealt with in a manner arranged by him with the Engineer.

I examined these columns carefully and I found that, so far as can be judged by sound from tapping the iron casing of the columns, and by inspection at the peep holes and other holes which have been made in the casing, the concrete has now been made good, all the hollow places having been filled up with cement grouting.

Orders have been given to continue this work on all the columns, but until all have been satisfactorily healed in the same manner as those which I have tested, I cannot recommend that the opening of this line should be sanctioned and I must report that by reason of the incompleteness of the works, it cannot be opened without danger to the public using it.

I have, etc.,
F. A. Marindin
Major

Report of Maj. Gen. Hutchinson. 3rd. November, 1883.

Sir,
I have the honour to report for the information of the Board of Trade, that, in compliance with the instructions contained in your minute of the 25th ultimo, I have revisited the Whitby, Redcar, and Middlesbrough Union Railway.

Since Major Marindin's report [see #2] re-inspection of this line in August last, I find that the concrete in the columns of the viaducts has been carefully gone over, and its defects made good, except in three places in Numbers 5, 6, and 8 piers at Staithes, which were to be at once attended to.

I made a further examination of the condition of the cliff at the first bay north of Sandsend station, and am of the opinion that it is desirable at once to build a sea wall at its foot as a precautionary measure, to this Mr. Harrison, the North Eastern Company's Chief Engineer, has consented.

I have also arranged with him that considering the great height of the Staithes viaduct and its exposure to easterly gales, a wind gauge shall be placed in a suitable position in charge of the Staithes station master, and that no train shall be allowed to cross the viaduct when the wind registers a force of 28 lbs to the square foot or more.

Owing to the peculiar construction adopted for the viaducts it will be desirable that the speed in running over them shall not be allowed to exceed twenty miles an hour.

The viaducts will require most careful maintenance particularly as regards the bracing, both longitudinal and transverse.

The line between the tunnels next to Sandsend station and that station will require very careful watching on account of the dangerous character of the adjacent cliffs.

No time should be lost in painting those portions of wrought iron girders which have not yet been recently painted.

A temporary facing point south of Loftus station has to be removed before the line is opened for passenger traffic.

Subject to the foregoing remarks, and to the undertaking as to the working of the line, dated the 2nd instant, the opening of the Whitby, Redcar, and Middlesbrough Union Railway need not, I submit, be further objected to.

I have etc.,
[signed] C. S. Hutchinson
Major General.

BIBLIOGRAPHY

Primary sources

The National Archives, Kew

AN 82/17. Census of Passenger Trains week ending 27/3/55 and 3/9/55.
AN 82/83. N. E. Region passenger train censuses: Chief Officer's papers 1952-54.
AN 82/84. N. E. Region passenger train censuses: Chief Officer's papers 1954-55.
AN 82/85. N. E. Region passenger train censuses: General Manager's papers.
AN 82/86. N. E. Region passenger train censuses: Chief Officer's papers, n.d.
Rail 390/61. LNER Traffic Committee Minutes 1930.
Rail 390/1922. Report on the working of the Coast Line, Scarborough-Whitby-Middlesbrough. Summer 1934.
Rail 398/293. Station Traffic Book 1910-34.
Rail 398/427. Passenger Traffic Returns 1935-44 (actually 1935-7). N. E. area.
Rail 398/428. Passenger Traffic Returns 1939. N. E. area.
Rail 398/429. Passenger Traffic Returns 1940. N. E. area.
Rail 400/62. Passenger Traffic Returns 1938. N. E. area.
Rail 527/276. Whitby, Redcar, and Middlesbrough Union Railway; memoranda.
Rail 527/908. Plans for Deepgrove tunnel ventilating shafts (1900).
Rail 527/1088. Traffic arrangements with Whitby, Redcar, and Middlesbrough Union Railway.
Rail 527/1167. Traffic volumes, receipts, and expenses 1897-1907.
Rail 527/1748. Whitby, Redcar, and Middlesbrough Union Railway (1880-2).
Rail 743 (21 items): Whitby, Redcar, and Middlesbrough Union Railway:
  Rail 743/1. Minute Book.
  Rail 743/1. Minutes of the Directors' meeting, held at 24, Gresham Street, London, on 29th April, 1871.
  Rail 743/2. Minute Book (this includes a copy of the 1881 Prospectus).
  Rail 743/2. Minutes of the Directors' meeting held at 1, Victoria Street, Westminster, London, on 4th April, 1873.
  Rail 743/2. Minutes of the Directors' meeting held at 1, Victoria Street, Westminster, London on 11th July, 1873.
  Rail 743/2. Minutes of the Directors' meeting held at Old Palace Yard, London on 17th July, 1873.
  Rail 743/2. Minutes of the Directors' meeting held at (no venue given in the minutes) on 25th and 26th July, 1873
  Rail 743/2. Minutes of the Directors' meeting held at The Station Hotel, York on 21st. November, 1873.
  Rail 743/2. Minutes of the Directors' meeting, held at 7, Bank Buildings, Lothbury, London, on 6th. December, 1873.
  Rail 743/2. Directors' report to the half-yearly meeting of shareholders held at the City Terminus Hotel, Cannon Street, London, on 31st March, 1874.
  Rail 743/2. Directors' report to the half-yearly meeting of shareholders held at the City Terminus Hotel, Cannon Street, London, on 13th November, 1874.
Rail 743/3. Agenda Book.
Rail 743/4. Agenda Book.
Rail 743/7. Letters received.
Rail 743/9. Correspondence.

Rail 743/9. Letter of 30th November, 1873 [2].
Rail 743/9. Letter of 30th November, 1873 [1].
Rail 743/9. Letter of 2nd December, 1873.
Rail 743/9. Letter of unknown date, but probably 3rd December, 1873.
Rail 743/9. Letter of 4th December, 1873.
Rail 743/9. Letter of 5th December, 1873.
Rail 743/10. Correspondence.
Rail 743/12. Letters in/out.
Rail 743/12. Letter from the Joint Committee (of the WRMUR and the Scarborough and Whitby Railway) to the General Manager of the NER (23rd February, 1886).
Rail 743/12. Letter from the Audit Accountants Office (NER) to the WRMUR (13th April, 1886).
Rail 743/13. Letters in/out.
Rail 743/14. Weekly Traffic Receipts.
Rail 743/15. Engineering reports and land matters.
 Rail 743/15. Arthur S. Hamand's letter to the Chairman and Directors of the Whitby, Redcar, and Middlesbrough Union Railway of 5th May, 1874.
 Rail 743/15. Report submitted to the Ordinary Half-Yearly Meeting of Shareholders to be held at The Station Hotel, York, on Friday, the 24th day of October, 1879 at 2.30 in the afternoon.
 Rail 743/15. Report submitted to the Ordinary Half-Yearly Meeting of Shareholders to be held at the Company's Offices, 9, King's Arms Yard, London, E. C. on Saturday, 20th March 1880, at 12 o'clock noon.
 Rail 743/15. Report submitted to the Ordinary Half-Yearly Meeting of Shareholders to be held at the Company's Offices, 9, King's Arms Yard, London, E. C. on Wednesday, 13th October, 1880, at 3.30 p.m.
 Rail 743/15. Report submitted to the Ordinary Half-yearly Meeting of Shareholders to be held at the Company's Offices, 9, King's Arms Yard, London, E. C. on Thursday, 17th March, 1881 at 2.30 p.m.
 Rail 743/15. Report submitted to the Ordinary Half-Yearly Meeting of Shareholders to be held at the Company's Offices, 9, King's Arms Yard, London, E. C. on Tuesday, 3rd October, 1882.
Rail 743/16. Statistics and payments.
Rail 743/18. Cash Ledger Book.
Rail 743/20. Maps, Plans, and Elevations of Stations.
Rail 743/21. Correspondence and Reports re locos 'Penwyllt' and 'Mulgrave'.
Rail 943/27.
Rail 967/31.
Rail 1076/38. 1871 Prospectus of the WRMUR.
Rail 1110/501. WRMUR and NER correspondence.
 Rail 1110/501. Directors' report to the annual half-yearly meeting of shareholders held at The City Terminus Hotel, Cannon Street, London on 12th October, 1872.
 Rail 1110/501. Directors' report to the annual half-yearly meeting of shareholders held at The City Terminus Hotel, Cannon Street, London, on 4th March, 1873.
 Rail 1110/501. Directors' report to the half-yearly meeting of shareholders held at The Royal Hotel, Whitby, on 18th September, 1873.
MT 6. Board of Trade (Railway Department) 1851-1919.
MT 29. Railway Inspectorate; Inspectors' Reports 1840-1964 (104 vols.).

MT 29/44. (Inspectors reports of 7th July, 1883, 22nd August, 1883, and 3rd November, 1883.)

The Parliamentary Archives, The Palace of Westminster, London
HL/PO/PB/3/plan 1864/S13. (Scarborough, Whitby, and Staithes railway 1864. Plan, section, book of reference, published map, gazette notice, list of owners, lessees, and occupiers, estimate of expense).

HL/PO/PB/1/29&30 V1 n 250 [Local Act, 29&30 Victoria I, c. cxcv (1866)]. The deviation of line and alteration of levels of authorized line. Powers of N.E.R. Co., Amendment of Acts (HL/PO/PB/3/plan 1873/W13).

HL/PO/PB/1/1873/36 & 37 V1 n 162 (an Act to authorise the diversion and alteration of the line and levels of the WRMUR; and for other purposes) [Local Act, 36&37 Victoria I, c. cxxi (1873)].

HL/PO/PB/3/plan1873/W13 (WRMUR. Plan, section, book of reference, ordnance map, gazette notice, list of owners, lessees, and occupiers, estimate of expense).

National Rail Museum, York (search engine)
1983-8817

Newspapers
Whitby Gazette, 24th May, 1874. (Whitby Library).
Whitby Gazette, 5th August, 1927. (Whitby Library).
Whitby Gazette, 13th August, 1937. (Whitby Library).
Whitby Gazette, 13th September, 1957, p. 1. (British Library Newspapers, EW 1063).
Whitby Gazette, 20th September, 1957, p. 1. (British Library Newspapers, EW 1063).
Whitby Gazette, 27th September, 1957, p. 1. (British Library Newspapers, EW 1063).
Whitby Gazette, 11th October, 1957, p. 2. (British Library Newspapers, EW 1063).
Whitby Gazette, 20th November, 1957, p. 1. (British Library Newspapers, EW 1063).
Whitby Gazette, 21st February, 1958, p. 1. (British Library Newspapers, EW 1426).
Whitby Gazette, 7th March, 1958, p. 6. (British Library Newspapers, EW 1426).
Whitby Gazette, 14th March, 1958, p. 5. (British Library Newspapers, EW 1426).
Whitby Gazette, 2nd May, 1958, p. 1. (British Library Newspapers, EW 1426).
Whitby Gazette, 9th May, 1958, p. 1. (British Library Newspapers, EW 1426).

Secondary sources
D. H. Aldcroft, *British Railways in Transition, The economic problem of Britain's Railways since 1914* (London, 1968).
M. Bairstow, *Railways around Whitby, Volume 1*, (Halifax, 1989).
M. Bairstow, *Railways around Whitby, Volume 2*, (Halifax, 1996).
Bradshaw's August 1887 Railway Guide (Newton Abbot, 1968).
Bradshaw's April 1910 Railway Guide (Newton Abbot, 1968).
P. Butterfield, 'Grouping, Pooling, and Competition. The Passenger Policy of the London and North Eastern Railway 1923-39', *Journal of Transport History*, III ser., 7, 1986, 2, pp. 21-45.
S. Chapman, *Cleveland and Whitby*, Railway Memories No. 18 (Todmorden, 2007).
G. H. J. Daysh (ed.), *A Survey of Whitby and the Surrounding Area* (Windsor, 1958).
'Viaducts on the Whitby, Redcar and Middlesbrough Railway', *The Engineer*, 14th March, 1873, pp. 151, 158.

W. R. L. Forrest, 'Strengthening the East Row and Upgang Viaducts on the Whitby and Loftus Railway', *The Institution of Civil Engineers; with other selected and abstract papers* (ed. J. H. T. Tudsbery), Vol. CXXX, part 5, section 2, pp. 234-40.

T. R. Gourvish, *British Railways 1948-73. A Business History* (Cambridge, 1986).

K. Hoole, 'Footplate Farewell to the Whitby-Loftus line', *Trains Illustrated* (1958), Vol. XI, 119, pp. 410-15.

K. Hoole, *Railways in Cleveland* (Clapham, Yorks., 1971).

K. Hoole, *The Whitby, Redcar, and Middlesbrough Union Railway* (Nelson, 1981).

K. Hoole, 'Three Routes from Whitby to Middlesbrough', *Railway Magazine*, Vol. 104 (April 1958), 684, pp. 242-9.

H. L. Hopwood. 'From Saltburn to Scarborough by the Coast Line: North Eastern Railway', *Railway Magazine*, Vol. 47 (Nov. 1920), pp. 301-6.

E. M. Hutchinson, *Girder-making and the Practice of Bridge-Building in Wrought Iron* (London, 1879).

R. J. Irving, 'The Branch Line Problem in British Railway History. The Financial Evidence from North-East England', *Journal of Transport History*, III ser., 14, 1993, 1, pp. 27-45.

R. J. Irving, *The North Eastern Railway Company, 1870-1914, an economic history* (Leicester, 1976).

S. Joy, *The Train That Ran Away. The inside story of British Railways' chronic financial problems since nationalisation* (Shepperton, 1973).

T. Leunig, 'Time is Money: A Re-Assessment of the Passenger Social Savings from Victorian British Railways', *The Journal of Economic History*, 66, 3 (September, 2006), pp. 635-673.

J. Prebble, *The High Girders* (London, 1959).

P. W. B. Semmens, 'Scarborough-Whitby-Middlesbrough. The Coastal Line of the N. E. R.', *Trains Illustrated*, Vol. 4 (April, 1951), pp. 118, 123-5.

Official Guide to Whitby, (Whitby, 1934).

A. R. Thompson and K. Groundwater, *Cleveland and North Yorkshire (Part 2)*, British Railways Past and Present, No. 14 (new edition), (Kettering, 1994).

W. W. Tomlinson, *The North Eastern Railway; its rise and development* (London, 1914).

M. A. Williams, 'A Difficult Year in the History of the Whitby, Redcar and Middlesbrough Union Railway', *Journal of the Railway and Canal Historical Society*, 219 (March, 2014), pp. 32- 41.

M. A. Williams, '*A more spectacular example of a loss-making branch would be hard to find*': A financial history of the Whitby-Loftus line 1871-1958, unpublished University of York M. A. Thesis (2010).

M. A. Williams, 'Closing a line before Beeching: the end of the Whitby-Loftus line', *Journal of the Railway and Canal Historical Society*, 221, (November 2014), pp. 149-58.

M. A. Williams, 'The importance of fieldwork in researching railway history', *Journal of the Railway and Canal Historical Society*, 224 (November, 2015), pp. 377-387.

M. A. Williams, 'The Viaducts and Tunnels of the Whitby-Loftus Line', *Journal of the Railway and Canal Historical Society*, 218 (November, 2013), pp. 33-47.

M. A. Williams, 'The Whitby – Loftus line: "a more spectacular example of a loss-making branch would be hard to find." Is this really the case?' *Journal of the Railway and Canal Historical Society*, 216 (March, 2013), pp. 33-46.

M. A. Williams, *The Whitby – Loftus line, a history* (Saltburn, 2012).

C. Wolmar, *Fire and Steam. How the Railways transformed Britain* (London, 2007).

REFERENCES

CHAPTER ONE

1. R. J. Irving, 'The Branch Line Problem in British Railway History: the Financial Evidence from North-East England', *Journal of Transport History*, III ser. 14, 1993, pp. 2-7.
2. M. A. Williams, 'The Whitby-Loftus line: "a more spectacular example of a loss-making branch would be hard to find." Is this really the case?' *Journal of the Railway and Canal Historical Society*, 216 (March, 2013), pp. 33-46.
3. M. A. Williams, *The Whitby-Loftus line, a history* (Saltburn, 2012).

CHAPTER TWO

1. Tomlinson, W. W., *The North Eastern Railway: its rise and development* (London, 1914), pp. 469-70.
2. Ibid., pp. 611, 613.
3. Parliamentary Archives, Palace of Westminster, London, HL/PO/PB/3/plan 1864/S13.
4. To clarify: this is a 'Love Lane' in Scarborough, not the Whitby 'Love Lane'.
5. Ibid.
6. HL/PO/PB/3/ plan 1864/S13.
7. Parliamentary Archives, Palace of Westminster, London, HL/PO/PB/1/29&30 V1 n 250 [Local Act, 29&30 Victoria I, c. cxcv (1866)].
8. The deviation of line and alteration of levels of authorized line. Powers of N.E.R. Co., Amendment of Acts (HL/PO/PB/3/plan1873/W13).
9. HL/PO/PB/3/plan1873/W13. Also see page 38.

CHAPTER THREE

1. TNA RAIL 743/1. Minutes of the Directors' meeting, held at 24, Gresham Street, London, on 29th April, 1871.
2. In 1867 Arthur Hamand began practice on his own account in Birmingham, and, both alone and in conjunction with J. H. Tolmé, acted as engineer to various railway and tramway companies, among them the WRMUR.
3. TNA RAIL 743/15. This letter will be discussed at length later in this chapter.
4. TNA RAIL 527/276. For the full transcript of this vital document, see Appendix One.
5. TNA RAIL 1110/501. Directors' report to the annual half-yearly meeting of shareholders held at The City Terminus Hotel, Cannon Street, London on 12th October, 1872.
6. TNA RAIL 1110/501. Directors' report to the annual half-yearly meeting of shareholders held at The City Terminus Hotel, Cannon Street, London, on 4th March, 1873.
7. TNA RAIL 743/2. Minutes of the Directors' meeting held at 1, Victoria Street, Westminster, London, on 4th April, 1873.
8. TNA RAIL 743/2. Minutes of the Directors' meeting held at 1, Victoria Street, Westminster, London on 11th July, 1873.
9. TNA RAIL 743/2. Minutes of the Directors' meeting held at Old Palace Yard, London on 17th July, 1873.
10. TNA RAIL 743/2. Minutes of the Directors' meeting held at (no venue given in the minutes) on 25th and 26th July, 1873.
11. TNA RAIL 1110/501. Directors' report to the half-yearly meeting of shareholders held at The Royal Hotel, Whitby, on 18th September, 1873.
12. Ibid.
13. TNA RAIL 743/2. Minutes of the Directors' meeting held at The Station Hotel, York on 21st November, 1873.

14. TNA RAIL 743/9. Letter of 30th November, 1873 [2].
15. TNA RAIL 743/9. Letter of 30th November, 1873 [1].
16. TNA RAIL 743/9. Letter of 30th November, 1873 [2].
17. TNA RAIL 743/9. Letter of 2nd December, 1873.
18. Ibid.
19. Ibid.
20. TNA RAIL 743/9. Letter of unknown date, but probably 3rd December, 1873.
21. TNA RAIL 743/9. Letter of 2nd December, 1873. See page 79.
22. Ibid.
23. TNA RAIL 743/9. Letter of 5th December, 1873.
24. TNA RAIL 743/9. Letter of unknown date, but probably 3rd December, 1873.
25. See pages 43 and 44.
26. TNA RAIL 743/9. Letter of unknown date, but probably 3rd December, 1873.
27. Ibid.
28. TNA RAIL 743/9. Letter of 4th December, 1873.
29. TNA RAIL 743/2. Minutes of the Directors' meeting, held at 7, Bank Buildings, Lothbury, London, on 6th December, 1873.
30. TNA RAIL 743/9. Letter of 7th December, 1873.
31. TNA RAIL 743/9. Letter of unknown date, but probably either 13th or 14th December, 1873.
32. TNA RAIL 973/9. Letter of unknown date, but probably 3rd December, 1873.
33. TNA RAIL 743/9. Letter of 30th November, 1873.
34. TNA RAIL 967/31.
35. TNA RAIL 743/9. Letter of 5th December, 1873.
36. TNA RAIL 743/9. Letter of 2nd December, 1873.
37. See pp. 118-19.
38. TNA RAIL 743/15. Arthur S. Hamand's letter to the Chairman and Directors of the Whitby, Redcar, and Middlesbrough Union Railway of 5th May, 1874.
39. Ibid.
40. Ibid.
41. Ibid.
42. Ibid.
43. TNA RAIL 743/2. Directors' report to the half-yearly meeting of shareholders held at the City Terminus Hotel, Cannon Street, London, on 13th November, 1874.
44. How hopeless seemed the position of the company and, indeed, the future of the railway, is indicated by a report in the *Whitby Gazette* of 23rd May, 1874 which stated that 'all the horses recently employed in the construction of this line of railway between Whitby and Lofthouse, now suspended, were brought to the hammer (i.e. put up for sale) at Easington by Mr Thompson, auctioneer on Friday 15th inst. (i.e. May 15th). The 16 horses realized close upon £500, averaging about £31 each.' For details of this sale see page 52.
45. TNA RAIL 743/2. Directors' report to the half-yearly meeting of shareholders held at the City Terminus Hotel, Cannon Street, London, on 31st March, 1874.
46. Ibid.

CHAPTER FOUR

1. HL/PO/PB/3/plan 1864/S13. (Scarborough, Whitby, and Staithes railway 1864. Plan, section, book of reference, published map, gazette notice, list of owners, lessees, and occupiers, estimate of expense.)
2. HL/PO/PB/1/29&30 V1 n 250 [Local Act, 29&30 Victoria I, c. cxcv (1866).]
3. Daysh, G H J., (ed.), *A Survey of Whitby and the surrounding area* (Windsor, 1958).
4. Irving, ' Branch Line Problem', p. 42.
5. A contemporary representation of Upgang viaduct may be seen on page 62.
6. *Grace's Guide*. ' John Dixon, from the north-east of England and nephew of the John Dixon who was the first Chief Engineer of the Stockton and Darlington Railway, had moved to London in 1864, having considerable success abroad including the construction of a bridge over the Nile in Cairo and extensive drainage and sanitary works in Rio de Janeiro. He is best known for carrying out the transportation from Alexandria to London of Cleopatra's Needle.'
7. 'Viaducts on the Whitby, Redcar and Middlesbrough Railway', *The Engineer*, 14th March 1873, pp. 151, 158.
8. TNA RAIL 743 (21 items).
9. TNA RAIL 527/276; TNA RAIL 743/2; TNA RAIL 743/15.
10. See 1873 plan page 120.
11. *The Engineer*, 14th March, 1873, p. 158. Also see page 62 for illustrations.
12. TNA RAIL 743/2.
13. TNA RAIL 1110/501.
14. Ibid.
15. TNA RAIL 743/15.
16. TNA RAIL 743/9.
17. TNA RAIL 743/2.
18. TNA RAIL 743/9
19. TNA RAIL 1110/501.
20. TNA RAIL 743/2.
21. See page 81.
22. The following section is heavily dependent upon the article in *The Engineer* 14th March, 1873.
23. E. M. Hutchinson, *Girder-making and the practice of bridge building in wrought iron* (London, 1878).
24. TNA RAIL 527/276.
25. Ibid.
26. Prebble, J., *The High Girders* (London, 1959), pp. 69-71.
27. TNA MT 29/44.
28. Ibid.
29. Ibid.
30. Hoole, K., *Railways in Cleveland* (Clapham, Yorkshire), p. 46.
31. TNA RAIL 527/276.
32. TNA MT 29/44; Hoole, *Railways in Cleveland*, p. 44.
33. Williams, *Whitby-Loftus line*, pp. 71-3.
34. 'Strengthening the East Row and Upgang Viaducts on the Whitby and Loftus Railway'. *The Institution of Civil Engineers; with other selected and abstracted papers,* Vol. CXXX (Ed. J H T Tudsbery) (London, 1897), Part 5, Section 2, pp. 234-40. This article is of a highly technical nature.
35. Williams, *Whitby – Loftus line*, pp. 124-9.
36. TNA RAIL 743/2.
37. TNA RAIL 743/15.
38. TNA RAIL 743/2.
39. TNA RAIL 743.
40. HL/PO/PB/1/1873/36 & 37 V1 n 162 (an Act to authorize the diversion and alteration of the line and levels of the WRMUR; and for other purposes) [Local Act, 36&37 Victoria I, c. cxxi (1873)].
41. HL/PO/PB/3/plan1873/W13 (WRMUR. Plan, section, book of reference, ordnance map, gazette notice, list of owners, lessees, and occupiers, estimate of expense).

42. TNA RAIL 743/2.
43. TNA RAIL 743/15.
44. Tomlinson, W W., *The North Eastern Railway; its rise and development* (London, 1914), p. 669 'By an Act passed this session (1874) they (the WRMUR) were authorized to raise £100,000 of additional capital and to borrow a further sum of £33,000, but no one would take their shares or advance money on works of so costly a nature'.
45. TNA RAIL 743/2.
46. Ibid.
47. TNA RAIL 743/15.
48. Ibid.
49. TNA RAIL 527/276; Williams, *The Whitby-Loftus line*.
50. TNA RAIL 527/276.
51. Ibid.
52. TNA RAIL 743/15. Report submitted to the Ordinary Half-Yearly Meeting of Shareholders to be held at The Station Hotel, York, on Friday, the 24th day of October, 1879 at 2.30 in the afternoon.
53. TNA RAIL 743/15. Report submitted to the Ordinary Half-Yearly Meeting of Shareholders to be held at the Company's Offices, 9, King's Arms Yard, London, E. C. on Saturday, 20th March, 1880 at 12 o'clock noon.
54. TNA RAIL 743/15. Report submitted to the Ordinary Half-Yearly Meeting of Shareholders to be held at the Company's Offices, 9, King's Arms Yard, London, E. C. on Wednesday, 13th October, 1880, at 3.30 p.m.
55. TNA RAIL 1110/501. No photographs of the inside of Grinkle tunnel are available as the line is still in use.
56. TNA RAIL 743/15. Report submitted to the Ordinary Half-yearly Meeting of Shareholders to be held at the Company's Offices, 9, King's Arms Yard, London, E. C. on Thursday, 17th March, 1881 at 2.30 p.m. Also TNA RAIL 527/276.
57. TNA RAIL 527/276. Hoole, *Railways in Cleveland*, pp. 43-44.
58. TNA RAIL 743/15. Report submitted to the Ordinary Half-Yearly Meeting of Shareholders to be held at the Company's Offices, 9, King's Arms Yard, London, E. C. on Tuesday, 3rd October, 1882.
59. TNA MT 29/44 (Inspectors reports of 7th July, 1883, 22nd August, 1883, and 3rd November, 1883).
60. See page 77.
61. TNA RAIL 527/908.
62. Williams, *Whitby-Loftus line*, pp. 124-9.
63. Ibid.
64. See page 92.

CHAPTER FIVE

1. TNA, RAIL 527/1167.
2. TNA RAIL 398/293 (for 1910-34) and TNA RAIL 398/427 (for 1935-7).
3. TNA RAIL 400/62.
4. The United Bus Company timetables, many of which are given as appendices to this chapter, are deposited at the library of The Omnibus Society, 100-2, Sandwell Street, Walsall WS1 3EB. They do not have individual catalogue references.
5. Irving, 'Branch Line Problem', p. 36.
6. Leunig, T., 'Time is Money: A Re-Assessment of the Passenger Social Savings from Victorian Railways', *JER*, 66, 3 (September 2006), pp. 635-73.
7. Leunig, 'Time is Money', p. 637.
8. Ibid., p. 660.
9. Ibid., p. 669.
10. Ibid.
11. Ibid, p. 669.
12. TNA RAIL 743/20.
13. See *Appendix 2*.
14. For example, the Loftus receipts for 1905 show that passenger receipts

accounted for only 32 per cent of the total receipts for that year.
15. Irving, 'Branch Line Problem', p. 33. Irving gives a very clear definition of the term 'gross return'....[is] 'of course, a calculation of returns on investment before the cost of working the system, and hence bottom-line profit, is derived'
16. Ibid., 28; 34.
17. Ibid., Table 2, p. 34.
18. Irving, R. J., *The North Eastern Railway Company 1870-1914, an economic history* (Leicester, 1976), p. 187.
19. Ibid, pp. 189; 207.
20. Ibid, pp. 190; 205.
21. Leunig, 'Time is Money', p. 641.
22. Ibid, p. 661.
23. Ibid, p. 669.
24. Irving, *North Eastern Railway Company*, p. 42.
25. TNA RAIL 743/12. Letter from the Joint Committee (of the WRMUR and the Scarborough and Whitby Railway) to the General Manager of the NER (23rd February, 1886).
26. TNA RAIL 743/12. Letter from the Audit Accountants Office (NER) to the WRMUR (13th April, 1886).
27. *Bradshaw's August 1887 Railway Guide* (Newton Abbot, 1968), p. 296.
28. *Bradshaw's April 1910 Railway Guide* (Newton Abbot, 1968), p. 718.
29. TNA RAIL 398/293.
30. TNA RAIL 527/1167.
31. Irving, *North Eastern Railway Company*, pp. 228-9.
32. Ibid, p. 239.
33. Leunig, 'Time is Money, p. 669.
34. Irving, 'Branch Line Problem', p. 37.
35. Joy, S., *The Train That Ran Away*. The inside story of British Railways' chronic financial failures since Nationalisation (Shepperton, 1973), p. 115. This comment applies to a later period, but is nevertheless relevant for the period under discussion.
36. See graph page 7.
37. Aldcroft, D. H., *British Railways in Transition. The economic problems of Britain's Railways since 1914* (London, 1968), p.51.
38. Wolmar, C., *Fire and Steam. How the Railways transformed Britain* (London, 2007), p. 216. Wolmar argues that during the war 'many people were earning higher than normal wages with little to spend it on due to wartime shortages and were understandably eager to obtain some respite from the war. Passengers turned up in their droves at Christmas or Whitsun'. This behaviour could well have continued in the immediate post-war years.
39. www.metoffice.gov.uk/pub/data/weather/uk/climate/stationdata/durhamdata.txt. The Durham evidence is the closest, geographically, to Whitby, The Whitby records do not begin until 1961.
40. See graph page 98.
41. TNA RAIL 398/293.
42. These timings are taken from the 1938 summer timetable.
43. See graph page 99.
44. See graph page 101.
45. The best example of this wartime propaganda is the well-known poster by Bert Thomas in which a serviceman stands in front of a ticket office window in a station and demands, rather aggressively, if one's journey really is necessary.
46. TNA RAIL 398/428; TNA RAIL 398/429; TNA RAIL 400/62.

CHAPTER SIX

1. Forrest, W. R. L., 'Strengthening the East Row and Upgang Viaducts on the Whitby and Loftus Railway', *Proceedings of the Institution of Civil Engineers; with other selected and abstract papers* (Ed. J. H. T. Tudsbery), Vol. CXXX, 5, 2, pp. 234-40. Hopwood, H. L., 'From Saltburn to Scarborough by The Coast Line: North Eastern Railway', *Railway Magazine*, Vol. 47 (Nov. 1920), pp. 301-6. Hoole, K., 'Three Routes from Whitby to Middlesbrough', *Railway Magazine*, Vol. 104 (April, 1958), 684, pp. 242- 9. Idem, 'Footplate Farewell to the Whitby – Loftus line', *Trains Illustrated* (August, 1958), Vol. XI, 119, pp. 410-15. Semmens, P. W. B., 'Scarborough – Whitby – Middlesbrough. The Coastal Line of the N. E. R.', *Trains Illustrated*, Vol. 4 (April, 1951), pp. 118, 123-5. *The Engineer*, 14th March, 1873, p. 158.
2. Hoole, K., *The Whitby, Redcar, and Middlesbrough Union Railway* (Nelson, 1981); idem, *Railways in Cleveland* (Clapham, Yorkshire, 1971), pp. 40-52.
3. Four texts may be given as exemplars: Bairstow, M., *Railways Around Whitby* (Halifax, 1989); idem, *Railways Around Whitby, Volume Two* (Halifax, 1996); Chapman, S., *Railway Memories, 18, Cleveland and Whitby* (Todmorden, 2007); Thompson, A. R. and Groundwater, K., *British Railways Past and Present, 14, Cleveland and North Yorkshire* (Kettering, 1994).
4. Probably in 1957, but there is no direct evidence of the date of Cam Camwell's visit. Also, 10 years earlier, in 1947, the film *Holiday Camp*, directed by Ken Annakin was released in the U.K.. After casting around for a suitable location for the railway station shots in the opening sequence of the film, Sandsend station was chosen because of its unusual and beautiful position. In the sequence a train pulled by an 'A8' 4-6-2 tank (in LNER livery) draws into the station. Hundreds of eager holidaymakers on their way to the fictional holiday camp disembark. Such a scene had not been observed (if it ever had been) at Sandsend since the early 1920s.
5. Camwell, A., Vol. 2 (DVD).
6. http://www.youtube.com/watch?v=_RVxfqRLeGw
7. TNA MT 29/44 (Railway Inspectorate: Inspectors' reports [1883]).
8. Michael A. Williams, '*A more spectacular example of a loss-making branch would be hard to find*': A financial history of the Whitby-Loftus line 1871-1958, unpublished University of York M. A. Thesis (2010).
9. Williams, *The Whitby-Loftus line*.
10. Williams, 'The Whitby-Loftus line: "a more spectacular example of a loss-making branch would be hard to find". Is this really the case?', *Journal of the Railway and Canal Historical Society*, 216, March 2013, pp.33-46. Idem, 'The Viaducts and Tunnels of the Whitby-Loftus Line', *Journal of the Railway and Canal Historical Society*, 221, November 2013, pp. 33-47. Idem, 'A Difficult Year in the History of the Whitby, Redcar and Middlesbrough Union Railway', *Journal of the Railway and Canal Historical Society*, 221, March 2014, pp. 32- 41. Idem, 'Closing a line before Beeching: the end of the Whitby-Loftus line', *Journal of the Railway and Canal Historical Society*, 221, November 2014, pp. 149-58.
11. There are a number of different indicators for measuring past and present worth of money. It seems that £300 in 1873 was worth between £26,000-£33,000.

12. Williams, 'The Viaducts and Tunnels of the Whitby-Loftus Line', pp. 33-47.
13. HL/PO/PB/3/plan 1873/W13 (WRMUR. Plan, section, book of reference, ordnance map, gazette notice, list of owners, lessees, and occupiers, estimate of expense).
14. Memorandum of T. E. Harrison dated 14th November, 1883. There is a full transcript of this memorandum above, pages 164-166.
15. NRM York (search engine) 1983-8817.
16. *Journal of the Railway and Canal Historical Society*, Vol. 38, part 6, 224, (November 2015), pp. 381-82.
17. *Journal of the Railway and Canal Historical Society*, Vol. 38, part 7, No. 225 (March, 2016), p. 457.
18. *Whitby Gazette*, 13th August, 1937.
19. *Whitby Gazette*, 5th. August, 1927.

CHAPTER SEVEN

1. Many of these questions have been answered in a letter to the Editor of *The Gresley Observer* (The Gresley Society), 172 (Summer, 2017), p. 8. The letter's author quotes a further letter, by Peter Semmens, published in the August 1957 issue of the *Journal of the Stephenson Locomotive Society*: 'From the beginning of petrol rationing until 22nd February [1957], special trains were run every morning to take workmen from Whitby to Redcar special platform for Wilton. These men were mainly employed by contractors on the Imperial Chemical Industries' site at Wilton. These trains loaded to seven carriages, and Whitby was allocated two class L1 engines to work them, Nos. 67763/5. On the final timings the train left Whitby (Town) at 6.15 a.m., called at West Cliff and Sandsend, running non-stop from there to Redcar, arriving at 7.30 a.m. The locomotive and crew ran back light to Crag Hall and then worked the pick-up goods through to Whitby. The second crew worked the afternoon goods back to Crag Hall and then brought the return workmen's train from Redcar, leaving at 4.50 p.m. and reaching Whitby at 6.17 p.m. A second special train, consisting of a three-carriage push-pull set, hauled by a G5, ran out of Middlesbrough as far as Hinderwell in the morning and preceded the Whitby train.'

CHAPTER EIGHT

1. Williams, *The Whitby-Loftus Line*.
2. Aldcroft, D.H., *British Railways in Transition, The economic problem of Britain's Railways since 1914* (London, 1968), p. 116.
3. Gourvish, T.R., *1948-73. A Business British Railways History* (Cambridge, 1986), p. 66.
4. TNA AN 82/86.
5. TNA AN 82/83.
6. Ibid.
7. TNA RAIL 390/1922. Report on the working of the Coast Line, Scarborough-Whitby-Middlesbrough. Summer 1934.
8. Ibid.
9. Ibid.
10. See also Butterfield, P., 'Grouping, pooling, and competition. The passenger policy of the London and North Eastern Railway 1923-39', *Journal of Transport History*, III ser., 7, 1986, 2, pp. 21-45, at pp. 35-7.
11. Ibid. 206 Traffic returns for the seven stations on the Whitby-Loftus line for the years 1935-40 may be found at TNA RAIL 398/427, TNA RAIL 398/428, TNA RAIL 398/429 and TNA RAIL 400/62.
12. TNA RAIL 943/27.

13. Ibid.
14. Gourvish, *British Railways 1948-73*, p. 203.
15. Ibid, pp. 203-5.
16. TNA AN 82/86.
17. TNA AN 82/85.
18. *Whitby Gazette*, 13th September, 1957, p. 1 (British Library, EW 1063).
19. *Whitby Gazette*, 20th September, 1957, p. 1 (British Library, EW 1063).
20. *Whitby Gazette*, 27th September, 1957, p. 1 (British Library, EW 1063).
21. *Whitby Gazette*, 11th October, 1957, p. 2 (British Library, EW 1063).
22. *Whitby Gazette*, 20th November, 1957, p. 1 (British Library, EW 1063).
23. *Whitby Gazette*, 21st February, 1958, p. 1 (British Library, EW 1426).
24. *Whitby Gazette*, 7th March, 1958, p. 6 (British Library, EW 1426).
25. *Whitby Gazette*, 14th March, 1958, p. 5 (British Library, EW 1426).
26. *Whitby Gazette*, 2nd May, 1958, p. 1 (British Library, EW 1426).
27. *Whitby Gazette*, 9th May, 1958, p. 1 (British Library, EW 1426).
28. Irving, 'Branch Line Problem', pp. 27-45. T. Leunig, 'Time is Money: A Re-Assessment of the Passenger Social Savings from Victorian British Railways', *The Journal of Economic History*, 66, 3 (September, 2006), pp. 635-673.
29. Williams, "The Whitby – Loftus line: 'a more spectacular example of a loss-making branch would be hard to find.' Is this really the case?" , pp. 33-46.
30. Leunig, 'Time is Money', pp. 635-73.
31. Ibid., p. 637.
32. Ibid., p. 660.
33. Ibid., p. 669.
34. Ibid., p. 669.
35. Ibid., p. 669.

CHAPTER ELEVEN

1. TNA RAIL 1076/38, TNA RAIL 743/2.
2. *Official Guide to Whitby*, (Whitby, 1934), p. 124. The guide book provides a charming alternative, 'a very favourite plan is to take the train to Kettleness, the station before Hinderwell, and there alight. Cross the line, and ask the way to Runswick. If the tide be high, the cliff-tops must be taken; if low, then the most pleasant way is to descend to the rocks....'

Abandonment. The remains of Whitby (West Cliff) station.

N. Cholmondeley collection

INDEX

Where a page reference is in **bold** it is for an illustration.

1873 Deviation Act, 77
1876 Deviation Act, 82
1927 derailment, **122**, 129, 130, 131

A.P. Hunter, 144
Adits, 86, **87**
Arthur Hamand, 40, 42, 43, 44, 46, 48, 49, 50, 51, 59

Board of Trade, 5
Bog Hall Junction, **12**
Boulby Mine, 137

Camping coaches, 148

D.H. Aldcroft, 108
Debenture holders, 170
Deepgrove tunnel, 5, 8, 9, **18**, 19, 31, **75**, **86**, **89**, **91**, **92**, **129**, **130**, 137

E. W. Arkle, 139
E. W. Rostern, 139
Eastrow viaduct, **16**, **56**, **73**
Edward Corner, 46, 47, 49

F. Grundy, 144
F.M. Sutcliffe, 63, 125

George Fraser, 44, 46, 47, 48, 50, 51, 79
Gradients of the line, **30**
Grinkle, 5, 9, 10, 26, **27**, 100, **106**
Grinkle tunnel, 77, **78**, 83

Hawsker, 137
Henry Tennant, **124**, 125
Hinderwell, 5, 9, 21, **23**, **24**, 98, 99, 105, 106, **152**
Historiography, 115, 116
Holmsgrove, **126**
Horses, 52

Inland deviation, 60, **61**, 81

J.B.McClurg, 70, 146, 147, 148

J.H.Tolmé, 40, 42, 44, 58, 59, 60, 65, 74, 77
James Fraser, 44, 45
John Dickson, 39, 43, 44, 46, 48, 49, 50, 51, 58, 59, 65, 77, **165**
John Dixon, 57, 58, 60
John Waddell, 159, 164
Journal R&CHS, 7, 127

Keldhowe tunnel, 79, **80**
Kettleness, 5, 19, **20**, **21**, **22**, 31, **98**, **114**
Kettleness tunnel, 19, 20, 31, **76**, **84**, **90**

LNER, 95
Loftus, 5, 10, **28**, **29**, **31**, 39, **100**, **101**, **108**, **135**

Major Marindin, 67, 173, 174
Major-General Hutchinson, 66, 67, 68, 70, **171**, 172, 173, 174
Marchioness of Normanby, 39
Middlesbrough, 140, 141
Modernisation Plan 1955, 143, 144
Mr Eglinton, 46, 47, 48, 49
Mr Greenwood, 46, 48, 49

NER ticket pricing, 104
NER traffic returns 1897-1907, 102
NER/LNER traffic returns 1910-40, 108
Newholm viaduct, **15**, **55**, **74**
North Eastern Railway, 51, 79, 80, 81, 85, 154

Operating costs, 103, 107, 108
Operation of the line, 30
Overdale Beck viaduct, 120, 121
Owen's Cliff Signal Cabin, 123, 124, **128**

Prospect Hill Junction, **12, 13**, 14, 35, **138**, 157, 158, 159, **160**

R.J.Irving, 7, 94, 95, 102, 103, 107, 152
Railway plans: 1864, 33, 53
Railway plans: 1866, 34, 35, 36, 37, 54
Railway plans: 1873, 38

Railway: construction, 54
Railway: costs, 54, 85
Report on the working of the Coast Line 1934, 140, 141, 142
Robert Hodgson, 42, 43
Runswick, 21, 22
Runswick station, 161, 162, 163
Rural mobility, 109
Ruswarp, 35, 36, 37

Saltburn, 140
Sandsend, 5, 14, 15, **17**, **18**, 32, 97, 98, 109, 110, 111, 112, **117**, **126**, **136**, **150**
Sandsend viaduct, **56**, **72**
Scarborough, 141, 142
Skerne Ironworks, 57
Staithes, 5, 23, **25**, **26**, 32, 100, **151**
Staithes viaduct, 25, 31, **57**, 58, 59, **64**, 65, 68, **71**, 165, 166
Steeping Pits viaduct, 117, 120, **121**, **122**
'Sub' system, 46, 47

T. Leunig, 95, 96, 103, 107, 152, 153, 154
T.E.Harrison, 41, 42, 57, 65, 69, 82, 84, 85, **164**, 165

Tay Bridge, 5, 65
The Engineer, 58, 61
Transport Act 1947, 139
Transport Act 1953, 139
Transport Users' Consultative Committees, 143, 144

United Omnibus Co., 95, 112, 113
Upgang viaduct, 54, **55**, 58, **62**, 66, 165

Ventilation shafts, 86, **88**
Viaducts: cost, 60
Viaducts: delivery, 63
Viaducts: design, 61, **62**
Viaducts: destruction, 71

Whitby (Town), 5, 8, 10, **11**, 13, 32, 96, 97
Whitby (West Cliff), 5, 10, **14**, 97, 151
Whitby Gazette, 5, 146, 147, 148, 149, 169
WR&MU Rly, 39, 44, 104, 153
WR&MU Rly 1871 Prospectus, 161
WR&MU Rly 1881 Prospectus, 158, 161
WR&MU Rly Directors' minutes, 117
WR&MU Rly Engineer's certificates, 117, **118**, **119**

On the last day (3rd May, 1958) the penultimate train to Scarborough makes its way across the dramatic landscape between Kettleness and Deepgrove tunnels at Holmsgrove.

N. Cholmondeley collection